T0249764

Sustainable Management of Sediment Resources
Sediment Risk Management and Communication

The European Sediment Research Network SedNet was funded as a Thematic Network project (contract No. EVK1-CT2001-20002) by the 5th European Framework Programme for RTD, under the Key Action "Sustainable Management and Quality of Water" of the Environment Programme, topic 1.4.1 Abatement of Water Pollution from Contaminated Land, Landfills and Sediments. SedNet website: www.SedNet.org

Sustainable Management of Sediment Resources
Sediment Risk Management and Communication

EDITED BY

DR. SUSANNE HEISE

Environmental Science and Technology, Hamburg University of Technology, Hamburg, Germany

ELSEVIER

Amsterdam • Boston • Heidelberg • London • New York • Oxford
Paris • San Diego • San Francisco • Singapore • Sydney • Tokyo

Elsevier
Radarweg 29, PO Box 211, 1000 AE Amsterdam, The Netherlands
The Boulevard, Langford Lane, Kidlington, Oxford OX5 1GB, UK

First edition 2007

Notice
No responsibility is assumed by the publisher for any injury and/or damage to persons
or property as a matter of products liability, negligence or otherwise, or from any use
or operation of any methods, products, instructions or ideas contained in the material
herein. Because of rapid advances in the medical sciences, in particular, independent
verification of diagnoses and drug dosages should be made

Library of Congress Cataloging-in-Publication Data
A catalog record for this book is available from the Library of Congress

British Library Cataloguing in Publication Data
A catalogue record for this book is available from the British Library

ISBN-13:   978-0-444-51965-8
ISSN: 1872-1990

For information on all Elsevier publications
visit our website at books.elsevier.com

Transferred to Digital Print 2009
Printed and bound in Great Britain by
CPI Antony Rowe, Chippenham and Eastbourne

Working together to grow
libraries in developing countries

www.elsevier.com | www.bookaid.org | www.sabre.org

ELSEVIER   BOOK AID   Sabre Foundation
            International

Preface

A book on "Sediment Risk Management and Communication" invites a number of questions such as What are the risks both to and from sediments? How are they assessed? How should they be managed considering the European situation of having 72 transboundary rivers and an environmental law that requires a good ecological status of all surface waters by 2015? What role do communication and risk perception play in this story? Why should sediments be of any concern at all?

I believe that this book, written by a group of experts with different background in sediment management issues, from a variety of different European countries, representing academia, government and industry, can provide insights into these questions.

The issue of sediments is not new – scientists have been studying sediments for a long time. Navigation authorities must remove and dispose of sediment, since it has the annoying property of accumulating where it should not – e.g. in waterways, ports and harbours. But apart from those interested for scientific or navigation reasons, concern for sediments with regard to European environmental regulations, management and risk assessments has been relatively small, even though it has been known for decades that sediments have a strong capacity to bind contaminants (e.g. see Förstner & Müller in 1974 on heavy metals in rivers and lakes)[1]. In highly developed Europe, transport of contaminants to sediments via direct industrial emissions, urban waste water, technological accidents, mining, or diffuse sources was extensive from the start of the industrial revolution well into the last century. Sediments have been called "the long-term memory of water quality"; they potentially serve as sinks, but also as sources for contaminants. The term "chemical time bomb" was coined in the 1990s in order to get this message across.

Although water quality has improved due to extensive point source reductions in the last century, there is still the need to measure the impacts of historical pollution, for example heavy metal contamination in the Rio Tinto; PCB and HCH contamination in the Mersey estuary; HCB-contamination in the Rhine; and dioxin contamination in the Elbe river, to name a few. In some instances,

---

[1] Förstner U, Müller G (1974): Schwermetalle in Flüssen und Seen als Ausdruck der Umweltverschmutzung. Springer: Berlin Heidelberg. 225 pp.

the distances between identified sources and the sites in which the contaminants have accumulated and remain can be several hundred kilometres. However, when the European Water Framework Directive came into force in December 2000, the property of sediments as sources for contaminants was largely neglected.

Within the 5th Framework programme of the EC, however, the European Demand Driven Sediment Research Network (SedNet) was funded 2002 under key action 1.4.1., called *"Abatement of water pollution from contaminated land, landfills and sediments"*. SedNet then sought to extend the focus on surface waters in the Water Framework Directive to sediments and to stimulate discussion leading to a harmonised European Sustainable Management Approach for sediments. As well as a number of symposia and conferences, four working groups were formed within SedNet that soon became the pillars of the network. They were dealing with "Sediment Quality and Impact Assessment of Pollutants", "Sediment and Dredged Material Management", "Sediment Management at the River Basin Scale", and "Sediment Risk Management and Communication". The results of their workshops are published in this series under those titles.

Whilst the WFD has the shortcoming of not explicitly considering sediments as a potential source of contaminants, it represents an unprecedented achievement in another respect: For the first time, actions and measures to achieve a good ecological and chemical status of river waters are required on a river basin scale, in spite of numerous political and administrative boundaries between and even within European countries at this scale.

This ambitious task opens the door to sustainable environmental management that takes the transport of compounds along a river into account and links impacts to sources over considerable distances. While this is certainly necessary for the water phase, sediments have some additional properties which complicate their management. Public awareness of sediments is low, as they are mostly out of sight, and thus there is little public pressure or interest to drive sediment management decisions. Furthermore, sediment "polluters" are often difficult to identify (or may no longer exist) due to the persistence of the contamination and the accumulation of contaminant mixtures in the sediment, causing difficulties in the tracing, allocation and enforcement of responsibility for sediment contamination.

Hence, it was a challenge to form a working group to discuss the management and communication of sediment risk. Our aim from the beginning was to identify existing concepts on risk assessment, to analyse deficiencies in sediment risk management and to work on conceptual approaches in order to improve the way in which potential sediment impacts are recognized and dealt with. Thus, our task was not sediment management in technical terms, but rather it was the development of a concept for addressing the challenge of contaminated sediments with regard to their risk within the catchments and river stretches that they potentially affect.

During our first workshop we agreed on the following sediment risk management objective: *„to reduce risk posed by contaminated sediments to humans and ecological receptors to a level, deemed tolerable by society and to control and monitor sediment quality and ensure public communication with the final aim of complying with the EU WFD and habitats directive".*

In this respect, along with the scientific consideration of risk and how it should and could be managed, we were very much aware of the difficult challenge of different perceptions of risks and their communication towards stakeholder groups, including decision makers.

This group of people forming this working group was characterized by an enthusiasm for the topic and a fondness for creative discussions. They developed new ideas and concepts during and between four workshops. The results of these discussions, which are elaborated upon in this book, will hopefully initiate further studies and thoughts on sediment risk management and communication issues. We hope that this book represents a positive step forward towards sustainable management of sediments in our rivers.

At the moment, five years after the initiation of SedNet, we still have a way to go towards this aim. The European Commission still has not identified Sediment Quality criteria, even with the release of the Daughter Directive of 17 June 2006. However, changes are apparent. Due to continuous communication on national and international level, SedNet helped to demonstrate that water quality is influenced by historical contamination. Working groups have been formed by different institutions that discuss explicitly the management of contaminated sediments along rivers such as the Rhine and the Elbe. Studies have been carried out to assess risks due to contaminated sites along rivers. Workshops have been initiated and projects on public communication of sediment issues have been funded. Due to the Water Framework Directive, there is a growing awareness that stakeholders (local people, countries, administrations, user groups) may not only share a river but also the responsibilities, the benefits, problems, **and** the financial challenges in their catchment areas. This perception may offer a chance to manage risks connected to legacies of the past wisely and on a river basin level. The recent increase in frequency and severity of flood events that transport large loads of suspended material downstream underlines the urgency of such actions.

Future prospects

SedNet will continue to support the process of growing awareness among stakeholders. On a self-financed basis since EU-funding ended, the sediment network has followed its aim of integrating sediment issues and knowledge into European strategies to support the achievement of a good environmental status and of developing new tools for those goals.

When it started, SedNet was asked by the European Commission to stick to the key action under which it was funded, and consequently marine sediments and sediment quantity issues were assigned a minor role. In future, the approach to sediment (risk) management must also address these issues, which have been neglected before. With the European Marine Strategy in discussion and with the increasing influence of climate changes on the hydrological conditions in catchment areas, topics like soil-water-sediment interactions, sediment transport, diffuse pollution due to surface run offs and the impact of rivers on coastal systems will come into focus. We hope that this book will provide a basis for the discussions to come.

Susanne Heise, Hamburg, September 2006

Acknowledgements

This book is the product of four workshops of the working group "Risk Management and Communication", organized as part of the Thematic Network "SedNet". The work of this group was made possible through financial support from the European Commission under the 5th framework programme.

Its content is based on discussions carried out during and between the workshops and all people who joined one or more of the meetings contributed their arguments and comments to its outcomes. I would like to thank the following participants for the way they made long hours of work enjoyable and rewarding: Sabine E. Apitz, Eduardo Arevalo, Marc Babut, Helge Bergmann, Johan Bierkens, Jos Brils, Mario Carere, Claudio Carlon, Piet den Besten, Panos Diplas, Gerald Jan Ellen, Astrid Hadeler, Henner Hollert, Jan Joziasse, Alexandra Katsiri, Falk Krebs, Jaqueline Lavender, Vera Maaß, Amy Oen, Nico Pacini, Marina Pantazidou, Peter Simpson, Adriaan Slob, Tibor Haarosi, and Sue White.

Among them, I would like to especially and gratefully acknowledge the authors of this book for their enthusiastic commitment and for their cooperative efforts in its preparation. The writing meetings that we had will stay on in my memory due to their intensity and their friendly and creative atmosphere.

Special thanks also go to the native English speakers of our group for their language-aid and to Linda Northrup for revising the "European English" of the non-native speakers' chapters.

I am deeply indebted to Ulrike Meyer for formatting, Helge Bergmann for indexing and to Heinz Stichnothe for his everlasting willingness for discussion and constructive criticism.

Marc Eisma, Jos Brils and Wim Salomons served as peer reviewers for the whole volume. Chapter 3 has undergone a major review by Joe Jersak.

Jos Brils as the SedNet coordinator did a splendid job and qualified as trouble shooter. I thank the whole SedNet steering group for their support.

Susanne Heise
Coordinator of the SedNet working group on sediment risk management and communication

Contents

The Role of Risk Management and Communication in Sustainable Sediment Management

Susanne Heise[a] and Sabine Apitz[b,c]

[a]*Consulting Centre for Integrated Sediment Management (BIS), Technical University Hamburg Harburg, Eissendorfer Strasse 40, 21073 Hamburg, Germany*
[b]*SEA Environmental Decisions, Ltd. 1 South Cottages, The Ford, Little Hadham, Hertfordshire SG11 2AT, United Kingdom*
[c]*Institute of Water and Environment, Cranfield University, Silsoe, Bedfordshire, MK45 4DT, United Kingdom*

Sediments are an integral part of the aquatic ecosystem. They provide habitat for a large number of benthic micro- to macrofauna (organisms that live all their lives or during specific life cycles at the bottom of rivers, estuaries and marine waters). Major biogeochemical processes take place in sediments. As with ecosystem and hydrologic function in the water column, the functioning of sediments has been impaired by contaminants and other anthropogenic impacts over decades. Like river waters, sediments are transported, albeit often at slower rates, downstream, crossing political and regulatory boundaries. Hence, sustainable management of sediments requires a river basin approach.

Whilst water only reflects a transient signal of contaminants at a given time, sediments can accumulate and are able to store adsorbed contaminants for long time periods. These longer time scales in terms of contaminant retention and transport in the river basin can complicate the identification of polluters and thus the determination of responsibility and liability. Sediments can accumulate at slow and still water zones like reservoirs, harbours and bays, resulting in shallowing, loss of capacity and navigational barriers, thus causing a potential economic impact as well as posing potential human and ecological risks. However, unlike water, sediments are mostly invisible to the public, except in cases when their management directly concerns people (e.g., upland disposal sites). There is thus a need to raise public understanding of sediment issues and processes.

from Vogt et al 2003

Figure 1 European River Catchment areas [1]

From the river catchments in Europe (figure 1), 71 are shared by two or more countries [2], thus requiring joined efforts when quality needs to be improved. Although in Western Europe, quality of rivers like the Rhine has improved significantly since 1980, there is still cause for concern: the "Water and Wetland Index" of the World Wildlife Fund states that 50 out of 69 investigated river stretches (from 55 different rivers) in Europe are found to be of poor ecological quality, according to WWF criteria based on the impacts of canalisation, dams, pollution and altered flow regimes [3]. Worldwide, more than half of the major rivers are "seriously depleted and polluted, degrading and poisoning the surrounding ecosystems, threatening the health and livelihood of people who depend on them" (World Commission on Water 1999). Many of the organic and inorganic contaminants, that are emitted into the water phase have a tendency to partition from the dissolved phase to the active surface of fine-grained, organic-rich particles that form part of both bed and suspended sediments. Other contaminants already within or bound to the particulate phase enter the river system continuously. As such, while surface waters generally exhibit a transitory chemical signature indicative of current conditions, sediments can, in contrast, retain and integrate the chemical signature of decades to centuries (as can upland soils, waste deposits, agricultural and industrial sites and mining areas, all of which can ultimately influence sediments).

And even if water quality is good, poor sediment quality or hydrodynamically altered sediment conditions may threaten "the good ecological status" of a water body due to impacts on the benthic community, or, over time, due to biologically-induced, advective, diffusive and physical processes such as resuspension and flooding that may remobilise contaminants from the sediment matrix. Sediments are present in various depositional areas such as reservoirs, harbours, lakes, barrages and flood plains, and are susceptible to erosion and further transport downstream. Furthermore, continuing agricultural and industrial practices, as well as catastrophic spills, accidents, and changes in erosional and depositional patterns due to hydrological changes and anthropogenic activities continue to provide both point and diffuse sources of sediment, both contaminated and uncontaminated, into many river systems.

Sediment risk managers must address economic and societal, as well as environmental, risks. Acceptable levels of risk are determined by society, not only by science. Scientific risk assessment is an integral part of risk management; it is a tool that helps determine the probability and degree of risk and provides the scientific basis for decision making on management options. Such risk is generally defined as the product of the magnitude of a hazard and its probability. What risk management options are eventually selected, however, depend not only upon these scientific and technical issues, but also on legal, societal and economic considerations. These include societal perception of risk, which may be influenced by objective factors such as those above, but also subjective ones, including confidence in institutions, familiarity with issues, access to information and perceptions of control and potential personal impacts.

Thus, early, frequent and transparent communication between all parties (including the public) throughout the decision process is an essential part of risk management.

The implementation of the Water Framework Directive (WFD, 2000/60/EC) is changing the scope of water management from the local scale to basin (watershed or catchment) scale (often transboundary). It aims to establish a framework for the protection of inland surface, transitional (i.e. fjords, estuaries, and lagoons) and coastal waters and groundwaters in order to prevent further deterioration and protect and enhance the status of aquatic ecosystems and the water uses, as well as the terrestrial ecosystems and wetlands linked to them. The WFD approach encompasses measures of ecological, hydrological and hydrogeological systems including targets reflecting the ecological integrity of the water body. By 2015, Europe thus aims to achieve Good Ecological Status (GES) for its waters, measured against a reference derived from the *natural, unmodified conditions* for that water body type, rather than the status quo.

With the pragmatic view that certain areas will have to remain degraded or unnatural as the result of a developed landscape, water bodies which meet the strict criteria for designation as Heavily Modified Water Bodies (HMWB) or artificial water bodies (AWB), will only be required to have a Good Ecological Potential (GEP), rather than GES [4]. The GEP allows for morphological and ecological impacts of the physical modifications required by the continued use of the water body (e.g,. for flood defence, navigation, etc., [5]). Thus, the current process of water body designation has important implications for future activities such as dredging [5], flood control [6], construction, etc.

The WFD requires member states to develop River Basin Management Plans (RBMPs) which are designed to help to achieve the WFD objective of good, or improving, ecological status in all water bodies. These RBMPs will need to consider many aspects of basin-scale management within the socio-economic environment of the region, country and continent. However, it is of note that the WFD does not specifically address the important role of sediments in waters except as a habitat for macrozoobenthos, and even defines suspended solids as a pollutant, whether contaminated or not. There is, however, an indication that there will be more specific attention to sediments, as an integral part of the aquatic environment, in future European legislation. Historical contamination has already been recognized as a source for water contamination: By 2009, a programme of measures must be suggested against contaminating sources in river basins. This, logically, must include historically contaminated sediments, the successful management of which will be critical to the success of national and international basin management activities.

Sustainable sediment management along river basins will need to address the challenge of multiple and sometimes conflicting interests. Different stakeholders that need to be involved in the decision-making process, may, however, be driven by different environmental, socio-economic or legislative objectives. Representatives of NGOs may seek to preserve or improve the ecological quality of an area, harbour authorities may prefer the least costly approaches, the local public may be most concerned for their quality of life. The acceptance by all parties that proposed environmental, societal and/or economic risk management approaches are legitimate and are driven by different but often interlinked societal forces, is an important goal of risk communication. Identification of the various elements of the (perceived) situation such as the objectives, risks, indicators of risks and possible management options on both the basin and site-specific scale should facilitate the selection of measures or solutions, as well as helping to avoid alienation of stakeholder groups and the termination of communication (*Chapter 2: Sediment Management Objectives and Risk Indicators*).

Sustainable sediment management requires long-term solutions to sediment quality and quantity problems, which address potentially competing, but equally legitimate interests without creating new problems. It is thus necessary to define the interfaces between science, policy and economics, and to develop frameworks that allow decision makers to balance these issues. Such an approach will provide insight into changes in agricultural, industrial and development practices with highest potential to reduce sediment and contaminant inputs, and hence, cost of maintaining waterways and protecting the environment. Conceptual frameworks for sediment management that address the complexities inherent at both a basin-wide and site-specific scale must be developed and refined (*Chapter 3 – Strategic Frameworks for Managing Sediment Risk at the Basin and Site-Specific Scale*).

For both spatial scales, methods are needed for classifying management objectives and underlying risks in order to facilitate decision-making and to spend the scarce financial resources in an environmentally sustainable way. On a basin scale, prioritisation of sites along a river with regard to management objectives and involved risks will direct further risk assessment procedures to those areas where management options could have the highest positive impact on the river basin. Comparative risk assessment is needed if different management options need to be evaluated on a local, site-specific scale with regard to the management goals that are aspired (*Chapter 4 – Prioritisation at River Basin Scale, Risk Assessment at Local Scale*).

These management goals can be manifold: Dredged material may have to be disposed off and choices be made based on the involved risks amongst various options like relocation, disposal on land, treatment or beneficial use. The risk that contaminated sediments pose to the environment may need to be compared to the risk inherent with management options, e.g. remediation dredging. Potential risks include human exposure to contaminants via direct contact to sediments or through fish consumption, groundwater and surface water contamination, and degradation of environmental quality. The methods by which environmental risks are assessed, differ between countries. An overview of the different approaches to sediment assessment is given in Chapter 5 (*"Risk Assessment Approaches in European Countries"*).

Although European states differ in the priorities that they give to the sediment issue, environmental aspects of sediments are generally ignored in regulation unless dredging issues are concerned. This is possibly because of the uncertainties regarding political responsibilities and economic implications. Risk management, however, needs a regulatory context – on a national as well as an international level due to the large number of transboundary water courses. Chapter 6 (*Sediment Regulations and Monitoring Programmes in Europe*) reviews the status of sediments in various European regional, national and international statutes. Most attention is given to sediments when they accumulate in navigation areas, obstructing shipping activities, but their dredging and consequent handling is cost-intensive for responsible authorities. Relocation to the river or to the sea, or even beneficial use as fertiliser, beach or construction material have historically been regarded as the most economical disposal options on a short time scale, but in recent decades sediments have often been too contaminated to be relocated and thus must be treated or stored in specific terrestrial or sub-aquatic disposal sites. The ports of Rotterdam and Hamburg, for example, spend millions of Euros annually on dredging, handling, treatment and disposal of sediments.

As most of the contamination found in sediments derives from sources upstream, the removal of contaminated sediments from navigation areas downstream raises questions about who is responsible and liable for the pollution and who should pay for its removal and abatement. These issues can result in controversial discussions amongst stakeholders in river basins, which have the potential of becoming much more complex and contentious if they involve one of Europe's transboundary rivers. Managing contaminated sediments on a transboundary river basin scale may become one of Europe's largest environmental challenges because sediments reflect an industrial history for which responsible parties may no longer exist, either because a company's ownership or even the government (e.g., former East Germany) has changed, making the WFD's "polluter pays" principle a difficult prospect. The precautionary principle and the principles that pollution should be rectified at the source, the polluter should pay, and priority should be given to preventative actions are all enshrined in the European Treaty. Thus, the Environmental Liability Directive (ELD) holds polluters financially responsible for remediation of the effects of non-permitted emissions that happen after its implementation on 30 April 2007 (COM 2004) to baseline (pre-emission conditions). However, the ELD directive does not address the effects of historical emissions or the effects of diffuse emissions from permitted activities. It is unclear how this will be applied in regions with a legacy of historical pollution in sediments, or how issues of secondary pollution, such as the release of historically contaminated sediments during dredging activities, will be addressed.

The perception of responsibility for this contamination and the river quality depends on the position of the stakeholder. The managers of harbours who in some cases must deal with contaminated material from upstream, via dredging, treatment and/or disposal, may focus on the responsibilities of upstream polluters, though their own activities may also be affecting contaminant concentrations and mobility. Upstream stakeholders may perceive their individual contributions to the overall loads as minor, and they are thus not willing to accept partial responsibility for pollution of the whole river basin. As navigation authorities derive economic benefit as a result of dredging, some may argue that the costs they incur in handling sediments are the dredgers' problem regardless of contaminant source.

This discussion often goes largely unnoticed by the public. Since sediments are mostly hidden from sight, they are not generally the focus of people's attention. The impact of sediments on the environment, on social systems and on economies is often indirect. Public awareness is low: People who enjoy lying on beaches during summer and like to take a walk on the seaside find it difficult to acknowledge, that "mud" in rivers is essentially the same material as the sand on the beach – even though often of a finer grain size. This is confounded by the problem that material that needs to be dredged from the system is defined in many regulatory frameworks as "waste", not as a part of the ecosystem with an important ecological function [7]. The organic-rich material that originally fuelled the first great cities (for instance, the sediment brought up in the annual flooding of the Nile fertilized the great wealth of Egypt) is now seen as toxic, smelly and offensive.

There is often little correlation between risk as it is perceived by the public and risk as measured by scientific risk assessment. During the communication process and preferably at the beginning of a decision-making process, the risk perception of various stakeholders needs to be addressed. How a risk is perceived depends on individual character, and is influenced by cultural conditions and personal experiences. For instance, the perception of risk often increases when the public does not feel in control of a situation [8]. It follows that in sediment management the public feels more at risk from upstream emissions than from activities in their own region, and this perception will thus increase as one moves downstream. Different people involved in the decision making process may have widely differing attitudes to risk. These differences must be understood, respected and addressed if all stakeholders are to be successfully involved in the decision process. The WFD requires extensive stakeholder involvement at every stage of such a decision process. This puts great responsibility on the scientific community to better communicate the controversial role of sediments in economics, the ecosystem functions and the eventual success or failure of basin-scale sediment risk management (*Chapter 7 - Risk Perception and Communication*).

We hope that this book will help to increase awareness of the important role that sediment management must play if the environmental management of river basins is to be successful, and that it will facilitate discussions on a risk-based approach to sustainable sediment management.

REFERENCES

1. Vogt J, Colombo R, Paracchini ML, Jager Ad, Soille P (2003). CCM River and Catchment Database for Europe, Version 1.0 ed. EC-JRC
2. UNEP (2002): Global Environmental Outlook 3. Earthscan.
3. Hygum B, Madgwick J, Vanderbeeken M, Blincoe P, (Editors)(2001): WWF Water and Wetland Index Assessment of 16 European Countries – Phase 1 Results – April 2001. WWF Rep. Nr.: April 2001, 72 pp.
4. Hull SC, Freeman SM, Rogers SI, Ash JB, Elliott M (2004): Methodology for the Provisional Identification and Formal Designation of Heavily Modified Water Bodies in UK Transitional and Coastal Waters under the EC Water Framework Directive.
5. Brooke J (2004): The EU Water Framework Directive - A Wake Up Call.
6. Brooke J (2004): The EU Water Framework Directive - One Year On.
7. Salomons W, Brils J (Eds.). 2004. Contaminated Sediments in European River Basins. European Sediment Research Network SedNet. EC Contract No. EVKI-CT-2001-20002, Key Action 1.4.1 Abatement of Water Pollution from Contaminated Land, Landfills and Sediments. TNO Den Helder/The Netherlands
8. Asselt MBAv (2000): Perspectives on certainty and risk. The PRIMA approach to decision support. Kluwer Academic Publishers: Dordrecht.

Sustainable Management of Sediment Resources: Sediment Risk Management and Communication   9
Edited by Susanne Heise

Sediment Management Objectives and Risk Indicators

Jan Joziasse[a], Susanne Heise[b], Amy Oen[c], Gerald Jan Ellen[d], Lasse Gerrits[e]

[a]TNO Environment and Geosciences, Laan van Westenenk 501, Postbus 342, 7300 AH Apeldoorn, The Netherlands
[b]TU Hamburg Harburg, Eissendorfer Str. 40, D-21073 Hamburg, Germany
[c]NGI, P.O. Box 3930, Ullevaal Stadion, 0806 Oslo, Norway
[d]TNO Environment and Geosciences, Schoemakerstraat 97, Postbus 6030, 2600 JA Delft, The Netherlands
[e]Erasmus University Rotterdam, Department of Public Administration, Postbus 1738, 3000 DR Rotterdam, The Netherlands

1. Introduction

In this chapter, we wish to discuss different aspects of social and societal driving forces and objectives in sediment risk management and introduce indicators as triggers for selecting management options on a site-specific basis, as well as on a larger (eventually river basin) scale.

The guiding risk management objective that is followed in this book, is to "reduce risk posed by contaminated sediments to humans and ecological receptors to a level deemed tolerable by society and to control and monitor sediment quality and ensure public communication, with the final aim of complying with the European Water Framework Directive (WFD) and Habitats Directive". This comprehensive statement implies a number of factors:

- The principal issue of the management objective is risk reduction. This may include environmental as well as economical and societal risks.
- Controlling and monitoring are part of risk management. This includes the control of improvements due to management options as well as the assessment of current status.
- Desirable levels of risk are determined by society – this implies that, e.g. environmental protection limits may not be enforced if the stakeholders opt against it. On the other hand, decreased risk tolerance may require management actions despite there being no scientific necessity.
- Public communication and involvement is an essential part of risk management strategies, not only because this is legally required by various

European Conventions[1], but also because experience shows that risk tolerance decreases with limited access to information and with the feeling of being powerless and controlled by external forces. Resistance from stakeholders can prolong decision processes and increase costs and efforts in the long run, so that complications can only be prevented if they are involved in the decision process [1].

- The WFD and the Habitat Directive provide a legal framework for risk management objectives. Sediment issues are not given much attention and without that, the aims of these directives will not be achievable [2]. Hence, sediment as an integral part of the ecosystem will have to be managed within the scope of these environmental rulings.

Therefore, meeting regulatory criteria is one of four specific objectives relevant to sediment risk management (section 2), in addition to maintaining economic viability (section 3), ensuring environmental quality and development of the natural environment (section 4), and securing quality of human life (section 5). These objectives will be further explained later in this introduction section.

In the process of achieving these objectives, decisions must be made about acceptability of risks and when to intervene. The Scientific Advisory Committee of the German Federal Government on Global Environmental Change (WBGU) differentiated *normal, transitional* and *prohibited* risk areas based on the relationship of magnitude of hazard and the probability of exposure [3].

According to the WBGU (Figure 1), *normal* risk comprises low uncertainty with regard to the probability of exposure and hazardous effects, small hazardous effects, low probability of exposure, limited persistence of contaminants, reversibility of damage, low variability of probability and hazardous effects, and low social conflict potential. Daily risks that stakeholders face on a routine basis usually belong to this category of normal risk and are regulated – for the most part sufficiently – by existing legislation in Europe.

---

[1]
a) the Water Framework Directive
b) The Århus Convention (1998) gives the public the right to obtain information on the environment, the right to justice in environmental matters and the right to participate in decisions that affect the environment
c) Directive 2003/35/EC of the European Parliament and of the Council of 26 May 2003: provided for public communication with respect to the drawing up of certain plans and programmes relating to the environment and amending with regard to public participation and access to justice

Probability of Exposure

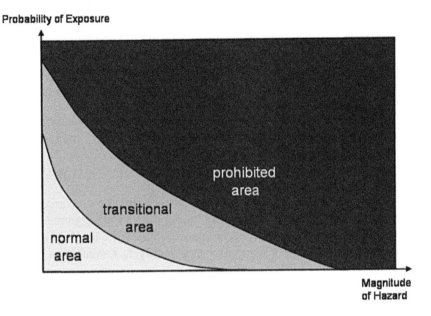

Figure 1: Normal, transitional and prohibited areas of risk (redrawn from WBGU (1999)

More difficult is *transitional* risk, which is characterised by the variability of highly uncertain exposure probability and an increased magnitude of hazard, increased persistence, ubiquity and irreversibility of effects, and high conflict potential, which may result in exodus, protests, resistance and refusal to cooperate on the part of persons concerned by these risks. These risks are difficult to manage, because the high uncertainty allows for different perceptions and interpretations of the magnitude of risk.

If the product of probability of exposure and magnitude of hazard grows too large compared with the expected benefits, these risks fall into the *prohibited* risk area and the decision that management actions are necessary, is easily taken (measures are necessary to circumvent the risks). An exception here is any risk that is known to be inherently hazardous and very likely to occur, but which will manifest itself only at some time in the future. These risks are often dismissed by stakeholders and decision makers. Examples are the effects of global warming, radioactive waste disposal, decline in biodiversity, increase of sea level, but also long-term effects of continuous accumulation of emissions into rivers.

The quantification of risks along such scales of increasing hazard and probability of exposure is highly complex. Risks comprise effects on sediment stability, the ecosystem and human health, they can be acute or chronic and follow different dose–effect relationships. Probabilities of exposure are weighted differently along a time scale and their evaluation depends on the kind of effects. With regard to extreme floods, a frequency of once a year is very high. When exposure to low concentrations of substances during swimming is a concern, once a year is probably not worth mentioning.

The highly multifarious phenomenon of risk can require that decisions on management options be made quickly. Indicators that are measurable and quantifiable provide the means to simplify a complex reality by integrating a variety of parameters and as such they can be useful for selecting among management options. Using indicators has become increasingly popular in the last decade. In order to help implement the action points of Agenda 21 following the Rio de Janeiro Earth Summit in 1992, the United Nations was asked to establish a set of 'indicators of sustainable development' that would help monitor the progress. A number of indicators addressing different environmental aspects have been developed and discussed since that time (e.g. [4-7]).

Environmental indicators are used for three main purposes in policy making [8]:
- to supply information, in order to enable policy-makers to value the gravity of environmental problems;
- to support policy development and priority setting by identifying key factors that put pressure on the environment;
- to monitor the effects of policy responses.

A statement reflecting the challenge inherent in choosing the right indicators has generally been accredited to Albert Einstein: '*Everything should be made as simple as possible, but not simpler*'. Slobodkin describes it in more words: ' ... all science is the study of either very small bits of reality or simplified surrogates for complex whole systems. How we simplify can be critical. Careless simplification leads to misleadingly simplistic conclusions' [9]. Risk indicators for sediment management must be chosen which are sufficient and adequate with regard to the management objective. Additionally, they must be accepted by stakeholders, and should allow risks to be ranked and sites to be prioritised.

Sections 2, 3, 4 and 5 in this chapter discuss the different risk management objectives. They are structured in such a way that they subsequently describe the social and societal driving forces, the specific risks involved, risk indicators and potential management options. At the end of each section a summarising table (tables 2, 3, 4 and 6) is presented. With respect to the risk indicators and management options a distinction is made between a site-specific and a river basin approach. Both water quality and water quantity issues are addressed. The tables aim to be as comprehensive as possible. The indicators could be used for a risk ranking of sediments for management purposes. The management options given, however, have different effects depending on the timescale and differ in value in terms of sustainability: Capping of contaminated sites (mentioned in table 3), for example, will have an almost immediate effect on contaminant transfer from sediment to the water column, exhibiting a positive effect on the water quality downstream, while source control and restructuring of the riparian zone may take longer, although these are long-lasting and highly sustainable options. In order to fulfil the management objectives on the river basin scale (called "Basin Management Objectives"), a combination of 1) effective, short-term, site-specific measures that have an impact on the larger river basin scale and 2) sustainable measures that will improve sediment and water quality over the long term, need to be addressed in a Basin Management Plan (see also Chapter 3 - *Strategic Frameworks for Managing Sediment Risk at Basin and Site-Specific Scale*).

Various management objectives and a definition of indicators of risk for those objectives, on both the basin and the site-specific scale, should facilitate the selection of measures or management options, as well as helping avoid alienation of specific stakeholder groups and the termination of communication.

Common understanding of the parameters involved and of their relation to each other are essential for communicating environmental decisions. Figure 2 depicts a potential concept about how these parameters interrelate:

Social or societal forces motivate, influence or otherwise drive sediment management. These include human values that have been documented in the form of directives and legislation. They comprise human needs that concern society, such as fishing, recreation and the ability to navigate on waterways for trade purposes. They also represent societal expectations and perceptions, such as the perception of risk and the willingness to preserve or improve quality of life. In practice, these attributes are represented by the various stakeholders (section 6).

Management objectives of risk management represent the overall aims that direct management options and are governed by the driving forces above. They comprise the need to meet regulatory criteria (section 2), maintain economic viability (section 3), ensure environmental quality and development of the natural environment (section 4), and secure quality of human life (section 5).

The objectives aim at avoiding or reducing **risks and (negative) impacts** to the environment, and this is done by implementing **management options,** if there is an indication that these risks are present (retrospective) or can be expected (proactive). Therefore, **risk indicators** are a necessary tool to connect risks with management options. Management options link back to the management objectives and risks/impacts and can address different scales: they can be applied on a river basin scale, on a sub-catchment scale, or on a local, site-specific scale. In this chapter, we only distinguish between the river-basin and the site-specific scale.

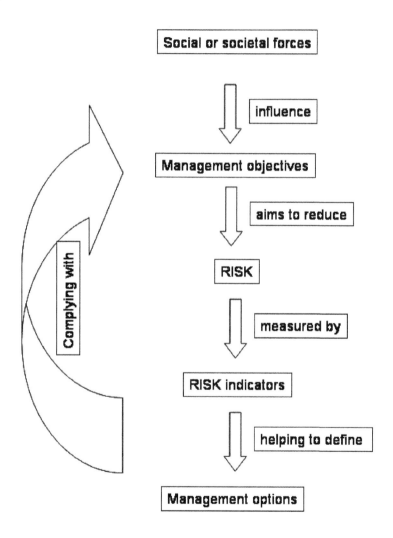

Figure 2: Structural approach of risk management parameters leading to management decisions

The type of management option and where to implement it should be selected after consideration of the site-specific situation and its effects on the river basin (See *Chapter 3: Strategic Frameworks for Managing Sediment Risk at Basin and Site-Specific Scale)*. The evaluation of site-specific and river basin-wide risks needs to be an interactive process, leading to site prioritisation and allowing for allocation of funds to those sites, with the highest expected effect on risk reduction in the river basin (see *Chapter 4: Prioritisation at River Basin Scale, Risk Assessment at Local Scale)*.

Table 1: Properties of site-specific and river basin (RB) risk indicators (RI)

Properties	Site-specific RI	RB-wide RI
Responsiveness to events	High	Low
Implications for river basin risk	Moderate	Very high
Risk perception of individuals	High	Low
Rate of data gathering	Fast and easy	Complex

Although site-specific and river basin (RB) oriented risk indicators can be the same (e.g. increase in flood events, decrease in fish yield), they are characterised by different properties (Table 1) and thus may lead to different effects (e.g. a merely local reduction in fish abundance without an effect on the whole river system). In some cases, indicators reflect the respective scales: e.g. negative public responses in opinion surveys (site-specific) and stakeholder movement (river basin scale).

2. Sediment Management Objective: Meeting Regulatory Criteria

2.1. Social and societal driving forces

Even though Europe is a small continent, it has the greatest number (71) of international river basins when compared with the other continents (Figure 3).

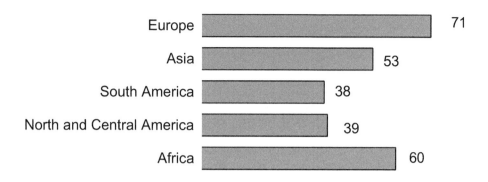

Figure 3: Number of international river basins [10]

It has become widely recognised that only management plans that address the scope of the river basin will improve environmental quality and sustainable use of water resources. This is the essential strategy of the European WFD and has already been established in the foundations of International River Commissions, such as the International Commission for the Protection of the Rhine that was founded in 1963. Since then, there have been many multilateral and bilateral agreements for the management of transboundary waters. These almost never mention sediments explicitly, even though the focus will have to turn to sediments when the objectives of the environmental regulations should be met. Regulations, directly or indirectly affecting sediment management are international conventions, EU water laws and other EU council directives.

International conventions:
Conventions (also known as treaties or agreements) as defined by the European Commission[2] are the most widely used instruments of public international law, which states that there is no general and systematic system of courts that can penalise a State in case it fails to comply with its obligations under an international convention.

The following three conventions aim to prevent negative impacts to the environment by disposal (relocation) of dredged material in estuarine, or marine areas: The **London Convention (LC, 1972)** aims to prevent marine pollution by dumping wastes and other matter worldwide. The **Oslo-Paris Convention (OSPAR,1992)** addresses the protection of the marine environment of the North-East Atlantic, and the **Helsinki Convention (Helcom, 1992)** deals with the protection of the marine environment of the Baltic Sea Area.

---

[2] for definitions on legal terms used by the EC see http://europa.eu.int/comm/justice_home/ejn

The **Ramsar Convention** (Convention of Wetlands of International Importance Especially on Waterfowl Habitat) was signed in the city of Ramsar, Iran, in 1971. It focuses on the conservation of waterfowl and their habitats, including issues of water quality, food production, and biodiversity, addressing all wetland areas, including saltwater coasts. Parties are obliged to list at least one wetland site of importance, establish nature reserves, make wise use of those sites, encourage biodiversity and supply information on implementation of policies related to the sites.

EU Water Laws (see also Chapter 6: *Sediment regulations and monitoring programmes in Europe*):
Directives are instruments of community law that are binding on the Member States with regard to the results to be attained. Once adopted under the EC Treaty by the European Parliament, the Council or the Commission, they have to be implemented by national law.
The Surface Water Directive (75/440/EEC) aims to protect and improve the quality of surface waters used for the abstraction of drinking water. It sets standards and requires member states to draw up a global and coherent plan of action for all waters, with a special focus on poor quality waters. It was designed to promote a reduction in pollution within 12 years of coming into force.
The Drinking Water Directive (80/778/EEC) establishes quality standards for drinking water and is a key instrument for safeguarding public health. These standards apply to a range of substances, properties and organisms. The directive is particularly strict with regard to microbiological parameters, given the public health implications of poor drinking water quality.
The Urban Wastewater Treatment Directive (91/271/EEC) addresses nutrient-based, bacterial and viral pollution caused by urban wastewater. Urban wastewater that discharges excessive levels of nutrients (in particular, phosphorus and nitrogen) into rivers and seas causes eutrophication, which leads to a decrease in oxygen levels and can finally be responsible for the death of most living organisms in that area. Introduction of potentially harmful bacteria and viruses by discharges also poses human health risks in waters that are used for bathing or aquacultures. The directive requires that urban centres meet minimum wastewater collection and treatment standards within specific deadlines, which are set according to the sensitivity of the receiving waters and the size of the affected urban population.
The Bathing Water Quality Directive (76/160/EEC) aims to ensure that bathing waters meet minimum quality criteria by establishing a set of binding and strict EU standards for a range of key parameters including indicators of the presence of faecal bacteria. The Directive requires that Member States carry out

regular water quality monitoring and send annual reports to the Commission, detailing bathing water quality.

The Groundwater Directive (80/68/EEC) requires Member States to apply a system of investigation and authorisation to waste disposal and other activities in order to ensure that groundwater is not polluted by dangerous substances.

The objective of the Fish Water Directive (78/659/EEC) is to protect and improve the quality of fresh waters that support, or could support, certain species of fish.

The Shellfish-Water Directive (79/923/EEC) requires Member States to designate waters that need protecting or improving in order to support shellfish. It also requires, among other things, that mandatory quality standards be achieved in the designated waters. It requires Member States to do regular sampling and to establish pollution reduction programmes.

The Dangerous Substances Directive (76/464/EEC) provides the framework for subsequent regulation to control the discharge of specific dangerous substances. The objectives are the elimination of pollution by the dangerous substances listed in the so-called Black List, and the reduction of substances listed on the Grey List.

The Water Framework Directive (2000/60/EC): The Water Framework Directive (WFD) is the most significant piece of European water legislation in more than 20 years. It will update most of the existing water legislation in Europe and for the first time addresses comprehensive river basin management with the legislative backing from all EU member states. It came into force when it was published in the Official Journal of the European Communities on December 22nd, 2000 as Directive **2000/60/EC** (OJ L 327/1). Its key objectives as set out in Article 1 are:

- to prevent further deterioration and to protect and enhance the status of aquatic ecosystems and associated wetlands;
- to promote sustainable water use based on long-term protection of available water resources;
- to aim at enhanced protection and improvement of the aquatic environment;
- to ensure the progressive reduction of groundwater pollution and to prevent its further pollution;
- to contribute to mitigating the effects of floods and droughts.

The member states are required to achieve a 'good chemical status' (referring to pollutants) for all water bodies (groundwater and surface water) and a 'good ecological status' of all natural surface waters by 2015. For heavily modified water bodies a 'good ecological potential' has to be achieved.

The WFD also supports establishing appropriate water pricing policies and participation of stakeholders in environmental decision making.

Groundwater Daughter Directive (COM 2003/550)

The proposal for the Groundwater Daughter Directive was adopted September 19th, 2003 (COM 2003/550) with the aim of setting out detailed provisions for the protection of groundwater and in reference to the objective of the WFD to achieve good chemical status by 2015. The proposed directive is intended to ensure that the monitoring and evaluation of groundwater quality is standardised across Europe

Other council directives:

A number of council directives do not specifically address the aqueous environment, but can still be of concern for sediment managers:

The Council Directive on the Conservation of Natural Habitats and of Wild Fauna and Flora (92/43/EEC) aims to contribute to the protection of biological diversity in the EU through the conservation of natural habitats and wild fauna and flora on the European territory of the member states by establishing a European ecological network of representative sites (known as Natura 2000) and ensuring that selected habitats and species are maintained and protected in order to maintain and/or restore them to a 'favourable conservation status'. Special Areas of Conservation (SAC) are to be designated in order to ensure habitat and species protection.

The Wild Birds Directive (79/409/EEC) requires member states to protect naturally occurring wild birds, especially those species that are listed in Annexes I–V of the directive, and their habitats by designating and managing Special Protection Areas and prohibiting certain harmful activities. This involves taking special conservation measures to ensure that wild birds and their habitats are protected and that the populations of all wild birds are maintained at levels which correspond to ecological, scientific, and cultural requirements. Member states are required to take measures to ensure that these objectives are met, including preserving, maintaining, and re-creating habitats.

The Environmental Impact Assessment (97/11/EC). This directive is one of the Community's principal pieces of environmental legislation. The aim of an Environmental Impact Assessment (EIA) is to identify and describe the environmental impacts of projects and to assess them with a view to their possible prevention or mitigation. The directive requires member states to carry out an EIA for certain public and private projects before they are authorised, where they are likely to have significant effects on the environment. For some projects (such as motorway construction) listed in Annex I to the Directive, such assessments are obligatory. For others (such as urban development projects or peat extraction projects) listed in Annex II, member states must conduct a method of screening to determine which projects require assessment. They can apply thresholds or criteria, carry out case-by-case examination, or do a combination of these.

During the EIA procedure, stakeholders can contribute their knowledge and express their environmental concerns with regard to the project. The results of the consultation have to be taken into account in the authorisation procedure.

The Strategic Environmental Assessment Directive 2001/42/EC
The purpose of the SEA Directive on Strategic Environmental Assessment is to ensure that environmental consequences of plans and programmes which are prepared for agriculture, forestry, fisheries, energy, industry including mining, transport, regional development, waste management, water management, telecommunication, tourism, town and country planning or land use are to be assessed under national legislation during their preparation and before their adoption. The planning procedures are to be communicated to public and environmental authorities such that their opinions are considered in the planning procedure. Once a decision is made, stakeholders must be informed about it.

An important addition addresses the occurrence of transboundary effects:
'In the case of probable transboundary significant effects, the affected Member State and its citizens are informed and have the possibility to make comments which are also integrated into the national decision making process.'

The Council Directive on Waste (75/442/EEC) requires that waste be recovered or disposed of without endangering human health and without creating risks for water, air, or soil, nor for the fauna, flora, landscapes, and sites of particular interest. 'Waste' refers to any substance or object which the holder disposes of, or is required to dispose of pursuant to the provisions of national law in force; 'disposal' refers to: 1) the collection, sorting, transport and treatment of waste as well as its storage and tipping above or under ground, and 2) the transformation operations necessary for its re-use, recovery or recycling. For the implications of the Council Directive on Waste on sediment management, see Chapter 6 - *Sediment Regulations and Monitoring Programmes in Europe*.

2.2. The risk involved

2.2.1. The risk of getting a bad reputation

Public international law has no court overriding national decisions, if a state decides not to act according to an international **agreement** (as opposed to a directive or regulation) that it had previously signed. Infringements – whether assumed or real – of environmental conventions are often noted and made public by non-governmental organisations, potentially threatening the reputation of the state in question. A recent example can be found in Germany, which is one of three countries (after Belgium in the 1980s and Australia in 1997) that have invoked the 'urgent national interest' clause to restrict the boundaries of one of the 1420 Ramsar sites that were designated in February 2005. In 2000, 20% of the largest freshwater tidal ecosystem in the EU, the Mühlenberger Loch near Hamburg, with a rich primary and secondary production and a high level of fish and bird species diversity, was to be replaced by a new Airbus Industry factory, which would create new jobs in that area. This instigated protest by national (NABU[3] and BUND[4]) and international associations (such as IFAW, the International Fund for Animal Welfare; WWF, the World Wide Fund for Nature; and Greenpeace), raising public hostility against the proposal (although some of the public - being employed by Airbus - did not oppose the plans).

However, this roughly 700-ha tidal area was not only designated as a Ramsar Site in 1992, but also a Special Protection Area (SPA) according to the requirements of the European Union Wild Birds Directive (79/409/EEC) in 1997, and as a Site of Community Importance (SCI) according to the Habitats Directive (92/43/EEC) in the same year. Therefore this case did not only fall under public international law but also under community law and therefore required the official agreement of the European Commission that 'imperative reasons of overriding public interest' outweighed the adverse environmental effects of allowing Airbus Industry to extend the plant. In addition, the Habitats Directive (Article 6.4) required and the Ramsar Convention suggested (Article 4.2) that compensation measures be proposed.

This example shows how public perception influences the risk of getting a bad reputation. Notwithstanding all efforts, the public has mixed feelings about the extension of the plant.

2.2.2. The risk of being fined

Non-compliance with EU environmental laws may result in substantial fines. According to Article 226 of the Treaty, the Commission has the power to take legal action against a member state that is not respecting its obligation. Letters of Formal Notice are addressed to those member states which the Commission considers has committed an infringement of Community law that warrants the opening of an infringement procedure. In this *first written warning*, the member state concerned is requested to submit its observations by a specified date, usually within two months. Depending on the character of the reply or its absence, the Commission may decide to address a Reasoned Opinion (or *final written warning*) to the member state, which specifies why it considers that there has been an infringement of Community law and asks for its compliance

---

[3] NABU – Naturschutzbund Deutschland

[4] BUND – German branch of Friends of the Earth, Bund für Umwelt und Naturschutz Deutschland (BUND), the largest German environmental protection association

within a specified period. If the member state fails to comply with the Reasoned Opinion, the Commission may decide to bring the case before the Court of Justice. Article 228 of the Treaty allows the Commission to ask the Court to impose a financial penalty on the member state concerned.

The number of first and second warnings that have been issued by the EU is a sign of the pressure that is applied to national governments and transferred to various stakeholders. For instance, the Commission has decided to take further legal action against France, Portugal, The Netherlands, Germany, Spain, Ireland, Belgium, and Greece for non-compliance with EU laws on Water Quality (ECCE Brussels Brief – Annexes, January 2004). The European Commission is currently taking legal action against Belgium, Italy, Spain and the UK to ensure they comply with EU EIA legislation[5].

With regard to the protection and conservation of European biodiversity, the European Commission is taking action against eight member states – Luxembourg, Belgium, Italy, Austria, Spain, Ireland, Greece and the UK – for failing to ensure sufficient protection of wild birds, habitats and species.

In the case of the Mühlenberger Loch, Germany convinced the European Commission that for socio-economic reasons it was justifiable to partially destroy the habitat and that the suggested compensation measures were sufficient (Letter from the European Commission from April 19th, 2000).

A number of former water quality directives will be repealed by the WFD in the coming years, for example the Surface Water Directive, the Fish Water and Shellfish Water Directive, the Dangerous Substance Directive, the Groundwater Directive, as the objectives of the WFD will lead to fulfilment of the objectives of these first-phase directives. Currently, the WFD mentions sediments only with regard to macrozoobenthos community monitoring and assessment. The influence of sediments on water quality is not considered ; the potential effects of their resuspension during specific hydrodynamic events (increase in current velocity, floods), and the subsequent remobilisation of accumulated contaminants. Although in many European countries, water quality has been improved over the last decades because of decreased industrial emissions and the construction of wastewater treatment plants, these improvements are not observed to the same extent in sediments.

The role that sediments play in the Flora and Fauna, the Habitats and the Wild Birds Directive comprises the protection of autochthonous species, mainly fish, but also the provision of food for non-aqueous species, such as birds. Wetlands as temporary habitats for migrating birds and permanent biotopes with specific communities are included in these directives. In addition to economic

---

[5] The current statistics on infringements can be looked up at
  http://europe.eu.int/comm/secretariat_general/sgb/droit_com/index_en.htm#infractions

impairment from penalties, non-compliance could lead to the loss of habitats and reduced biodiversity. This has a strong implication for the quality of human life. Nature is highly valued – especially by European society – as a retreat from urban life.

The Bathing Water and Drinking Water Directives aim to protect public safety and only apply to specific areas that serve as bathing sites or drinking-water production sites. A focus here lies on opportunistic pathogenic micro-organisms that could lead to detrimental effects on human health. Sediments and particles have been shown to provide an environment in which bacteria can survive much longer in a metabolically reduced state [11,12] and in which they are much more difficult to detect [13-16]. Accumulation of bacteria and viruses on sediment surfaces increases the risk of uptake of a substantial amount of pathogens during swimming if the sediment becomes resuspended. Resuscitation of viruses and bacteria from sediment samples has been shown to occur [17].

2.3. Indicators of risk

Indicators of risks that can be a consequence of not meeting regulatory criteria can be measured either on a specific location (site-specific) or they can be observed when assessing the whole river basin – depending on 1) the scope these regulations address and 2) the potential effects they may have beyond this scope.

Regulations that address specific sites are the Bathing Water Directive (bathing sites), the Drinking Water Directive (those areas that are used for production of drinking water), and the Habitats Directive. Non-compliance with the Bathing Water and Drinking Water Directives do not affect sediment management on the river basin scale. Indicators here are how much the threshold values exceed those given in the directives, but should also include sediments as a testing compartment.

Exceeding regulatory criteria is also an indicator for non-compliance with other management objectives; however, with respect to sediment, risk indicators on **a site-specific scale** should also include:

- the concentration of contaminants in sediments: due to the different behaviour of contaminants in sediments, this may lead to modifications in the choice of priority (dangerous) substances;
- ecotoxicological effects of sediments: inclusion of different exposure pathways will facilitate the risk assessment of contaminant transfer from sediments to water;
- alteration of benthic community: choosing indicator species that are specifically sensitive to environmental changes, risks to the biotope can be pointed out and monitored. Classification of risk could be achieved if these

indicators or additional species are ranked with regard to their essential biotope function.

On a **river basin scale**, the chosen indicators must convey information about a risk downstream and not only at a specific site. Contaminants in sediments, for example, will have to be assessed for their persistence during transport, their K_{ow} and the related 'wash-out' potential and their transport behaviour, e.g. the size class of sediment to which they preferentially adhere. To indicate risks for river locations downstream, sediment stability must be compared with current velocities in different hydrological situations.

Suspended matter is the most likely form in which resuspended sediment is transported, and therefore must be monitored regularly and in correspondence to extreme events, and contaminant concentration measured.

A reduction in fish species, which tend to migrate in rivers, is another strong indicator of declined river quality, such as the disappearance of salmon from most European rivers.

2.4. Management options

Management options **on a site-specific scale** are the following: (1) reduction of existing emissions of sediment-bound contaminants to the river, (2) reduction of exposure of organisms to the contaminants (which can comprise removing the contaminated sediments by excavation or dredging of the material, or capping the contaminated site), (3) reduction of effects of accumulated contaminants, e.g. by in-situ treatment, (4) changing the land use of the site (which applies mainly to recreational sites, or those areas which serve as drinking water production sites; if no measures are feasible, e.g. because they comprise a high risk themselves, or because a cost-benefit analysis proved that they would be too expensive, a site may be closed for a specific use), (5) evaluating the site in the scope of river basin management; this could affect any cost-benefit analyses, the socio-economic evaluation of the site, and potential measures that are taken upstream with consequences for the specific site.

Management options **on the river basin scale** would comprise the following: (1) enforcement of regulations such as infringements of European laws, which will positively affect the achievement of the regulatory criteria, (2) analyses of sediment transport and application of suspended particulate matter (SPM)-transport models or schemes for the river basin in order to assess potential impacts along the river; this could lead to measures such as the construction of sedimentation basins aimed at improvement of downstream water quality, (3) negotiations with upstream and downstream stakeholders, which should lead to an increased knowledge of perceived risks and potential measures that can be carried out, (4) facilitated communication with stakeholders, which will take place either individually or together with NGOs that take part in the environmental decision-making process.

An overview of the risk indicators and management options that have been discussed in this section is given in table 2.

Table 2: Risk indicators and management options for the objective of meeting regulatory criteria

Example driving forces	Risk and impacts involved	Indicator of risk		Management option	
		Site-specific	River basin approach	Site-specific	River basin approach
European ecological goals implemented via: • Habitats Directive • Wild Birds Directive • WFD • Shellfish & bathing waters Directive • International treaties • National legislation	Failure to meet ecological goals Economic impact (penalties and claims) Political tension (intra- and international) Loss of public confidence Stakeholder activism	Non-compliance with directives, e.g. • High contaminant load • Eco-toxicological effects • Alteration of benthic community • Eutrophication • High number of *E. coli* or pathogens	Non-compliance with directives, e.g. • Indicators for habitat losses • High contamination of waters and sediment • Poor chemical and physico-chemical quality • Reduction in migrating species	In-situ treatment Capping Function change Turn to river basin management (source control) Sedimentation basins, dredging, excavation	Negotiations with upstream / downstream stakeholders Make use of / establish river commissions [Rhine, Elbe, etc.] Revision of industrial or agricultural policies Enforcement of regulations Exchange of knowledge and technology Enhancement of public communication and stakeholder involvement

3. Sediment Management Objective: Maintaining Economic Viability

3.1. Social and societal driving forces

The main aspect of social and societal driving forces that govern the management objective of maintaining economic viability is financial gain or loss. Very often these driving forces have a strong positive or negative relationship with those for the management objectives of meeting regulatory criteria (section 2), ensuring environmental quality and nature development (section 4) and securing quality of human life (section 5). The reason for this relationship is that they have economic consequences that can either directly or indirectly be expressed in terms of monetary cost or benefit.

Examples of societal forces that have direct and obvious cost consequences are put forth by organisations that have an interest in maintaining or sustaining navigation and fishery activities. E.g. with respect to navigation (transport of people, bulk material and goods) they often antagonise the objectives of those discussed in section 4 (Environmental Quality); it is a challenge to find acceptable compromises (see also under management options). Social and societal driving forces with respect to fishery on the other hand, represent interests that blend well with those mentioned in the other management objectives (sections 2, 4, 5), because sustainable fishery benefits from healthy fish populations and these – in turn - will benefit from a high priority given to good water and sediment quality, improving nature development and quality of life. Short-term fishery interests however, can be opposed to long-term interests (for instance the banning of intensive cockle fishery in the Dutch Waddensea, or the establishment of fish quotas).

For other driving forces the consequences for failing to meet the management objective cannot always be so easily translated into an amount of money gained or lost. This is especially the case when subjective interests or values have to be weighted against each other. In these cases valuation methods can be used that are applicable when budgets are restricted, which is a common situation, and priorities must be assigned. An example of such a method is the contingent valuation method (CVM), see section 3.3. As a rule, these social and societal driving forces are dealt with in sections 4 and 5, but as usual, there are also intermediate cases (objectives) that are more difficult to classify, or that encompass a number of different aspects, some of a more economic nature, others of a more ecological nature. An example of such a complex objective is the mitigation of flooding (either catastrophic events, or situations simply causing inconvenience during periods of heavy local rainfall, or high annual river discharge), or the mitigation of the opposite extreme event: long periods of drought, or very low river discharge. The main social and societal driving forces here stem from quality-of-life interests (safety, convenience), ecological interests (biodiversity and nature development, related to the potential management options; see below) as well as economic interests (agriculture, environmental planning, compensation for reduced operating profits, production loss from a shortage of cooling water). Another example of a complex management objective is the development of recreation facilities. Obviously, this has much in common with economic aspects (the tourism industry), but also with nature development and quality of life, because usually, recreation can flourish only in attractive areas that have a well-developed, recreational and interesting natural environment.

3.2. Risks involved

Risks are defined in a broad sense here. This means that not only risks for ecology, human health and spreading of contaminants, but also risks for economic loss, reduced attraction of recreational areas, reduced well-being, unemployment, etc. Waterways have to be maintained for navigation and water transport purposes. Decreasing depths of waterways caused by sedimentation lead to economic losses, so related risk and impacts are mainly financial and economic in nature. For instance, very often, dredging operations are delayed due to problems experienced in utilising or storing the contaminated material brought up in dredging operations. This results in a situation where ships can no longer be loaded to their full capacity, or where discharge capacity is no longer sufficient to prevent the occurrence of flooding during periods of heavy rainfall. The loss of an activity such as fishing or agriculture also has large financial and economic effects, but in addition to these, ecological, public health, employability and other sociological effects are important as well. If the ecological quality of river water or sediment is impaired, the likely result is a decline in the abundance and diversity of fish species and hence fishery yields. For agriculture, a decline in crop yields can be caused by the impaired soil quality that results when contaminated material settles on floodplains in periods of high river discharge. This may also lead to risks for public health as a result of elevated contaminant concentrations in crops, milk, or meat.

As stated in section 3.1 above, flooding has many repercussions, involving various types of risk. People may no longer feel safe in their homes, or have to deal with the inconvenience and costs caused by flooding. With a focus on economic aspects, it is obvious that industry and agriculture may be faced with costs resulting from production loss. Other problems, such as food contamination, have indirect economic consequences. Agricultural or industrial companies may be less willing to settle in certain locations assigned as retention or overflow areas, which may cause a rise in unemployment figures and harm development of the area.

An additional economic factor is the tourism revenue. Wetlands and nature reserve areas can contribute to this factor to a certain extent. The attraction of an area for public recreation profits from a varied and healthy natural environment and wildlife and from good ecological quality in general. If this quality is endangered, for instance as a result of pollution, this may harm the attraction and therefore tourism revenue. Often these interests are endangered by those of other stakeholders, such as human settlement (development companies etc.), but sometimes it is entirely possible to combine the various interests. Examples will be given in section 3.4.

3.3. Risk Indicators

3.3.1. Valuation methods

In some cases there is not really a difference between site-specific risk indicators and river basin scale risk indicators. To move from site-specific to river basin risk indicators is then a matter of up-scaling. This is especially true for indicators that can be expressed in monetary units. This type of indicator involves priced effects, i.e. effects that can be assigned a price based on market value. However, the valuation of unpriced effects is more difficult. To this end, the total economic value is often subdivided into actual use values and non-use values; see Fig. 3 [18,19].

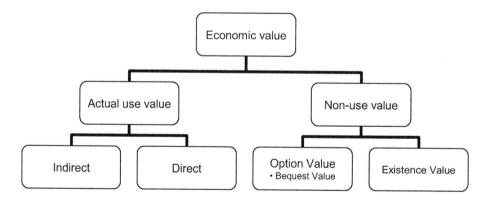

Figure 3: The economic value of ecosystems

The actual use value comprises both direct use of the environment (e.g. fish for consumption, trees for timber or paper production, water for drinking, irrigation, recreation) and indirect use of the environment (e.g. natural retention capacity, clean drinking water provision, oxygen production of forests). The non-use values of an environment include the option value and existence value. The option value is defined as the value to humans to preserve an environment as a potential benefit for themselves or others in the future (e.g. preservation of tropical rainforests as a potential source for medicines in the future). One specific type of option value is the bequest value, the desire to preserve an environment for future generations. The existence value is the value to humans for the simple existence of ecosystems (the existence nature's value) and is reflected – among others – in the willingness of people to pay for preserving environmental functions even when they themselves will never be able to benefit directly from them (the lifeline of organisations such as Save the Whales). It is the intrinsic value of ecosystems to humans only, and does not include the intrinsic value residing in the ecosystems themselves. See also Section 4 for a discussion on these issues.

If market prices are not available, or distorted, shadow prices can be calculated to estimate the economic value of environmental goods. Four methods can be distinguished:
1. The contingent value method (CVM). This method yields the consumers' willingness to pay (WTP) for a hypothetical change or willingness to accept (WTA) compensation for that change.
2. The hedonic pricing method (HP). This method utilizes statistical techniques to identify values of environmental 'goods' (e.g. the pollution level) as a function of marketed goods (e.g. house prices).
3. The travel cost method (TC). This is the oldest non-market valuation method. It assigns a value to environmental goods by using travel costs made by consumers to obtain them. It could be considered as a specific type of HP method.
4. The production function approach (PF). This approach links environmental quality changes to changes in production relationships. Two types of this method can be distinguished: (a) averting expenditure, based on the relationship between an environmental good and its perfect substitute (avoidance of costs for this substitute) and (b) dose-response, based on the relationship between an environmental good and a marketed good. In the PF approaches the difficulty lies in finding suitable substitutes or marketed goods with which a relationship is established.

Method 1 is a so-called stated preference (SP) method, making use of questionnaires that ask people to express the importance they attach to specific subjects. Methods 2-4 are revealed preference (RP) methods, taking peoples' behaviour in practice (markets) as a starting point. The latter methods are considered to be more reliable because the results are derived from actual behaviour of people [20].

The subject of economic valuation is too comprehensive to be further elaborated here. The interested reader is referred to the literature references mentioned above.

3.3.2. Indicators of site-specific risk

For the risk of economic losses from decreasing water depths (navigation and discharge capacity drivers) a site-specific risk indicators is the shallowing of waterways. For navigation the decrease in potential ship tonnage can be regarded as a derived risk indicator. Another site-specific risk indicator is the decrease in harbour activities (including e.g. ship-building activities) due to a reduced accessibility for ships. The risk and impacts of dredged material disposal have site-specific indicators in the storage capacity available, the costs (e.g. the storage price per m^3 or per ton of dry matter), or the failure to comply with regulatory requirements.

The decrease in fishery yield and the toxic loads in fish can be seen as indicators (site-specific and river basin scale, depending on fish migration characteristics) for the risk and impacts of the loss of fishery activities caused by pollution. Also, eutrophication and habitat change are site-specific indicators for this risk. For the risks involved with flooding, the risk indicators are the frequency of flood events, the rise in the average water level and the payment of insurance benefits. When flooding results in the loss of agricultural activities, additional risk indicators are the decrease in agriculture yield and toxic loads in crops, milk and meat.

In periods of extreme drought and low river discharge, economic consequences can be expected for shipping due to reduced water depths and for agriculture due to low groundwater tables and a shortage of water for irrigation. For power plants and other industrial plants, a shortage of cooling water may arise. Often this problem occurs in warm summer periods, when the maximum temperature of the surface water on which the cooling water is discharged, is exceeded. This may lead to a situation where the (electricity) production has to be reduced.

With respect to the risk of economic loss from a decline in tourism, a large number of site-specific driving forces can be named: the number of visitors or overnight stays, restaurant turnover, usage of yachting facilities, etc. In addition, indirect entities, such as eutrophication, noxious odours, negative publicity, or the loss of public awards for a clean environment ('Blue flag') can give an indication of the risk involved.

3.3.3. Indicators of risk at the river basin scale

For the risk of economic losses from decreasing water depths (navigation and discharge capacity driving forces) the river basin scale risk indicators are the same as the site-specific indicators. In addition, the use of alternative navigation routes could be seen as a river basin scale risk indicator. At this scale the risk of dredged material disposal reveals itself in a decrease in maintenance dredging activities. Furthermore, the volume limitations of disposal facilities and the loss of resources can also be regarded as risk indicators: The sediment stored in disposal facilities is removed from the fertiliser and construction material markets.

The river basin scale indicators for the risk of the loss of fishery activities are essentially the same as the site-specific indicators, although eutrophication is probably more properly marked as a site-specific rather than a basin scale indicator. Furthermore, for the risks involved with flooding and drought, the basin scale indicators are the same as the site-specific indicators and the same is true for the number of visitors or overnight stays, restaurant turnover, negative publicity, etc. as risks of economic loss due to a decline in tourism.

3.4. Management options

Whether a specific management option is regarded as site-specific or as a basin scale option depends on the scale of the measure, on the location along the river, but also on the driving force considered. Source control for instance, can be primarily site-specific if it is intended to decrease the noxious smell at a particular location, or in other cases where this measure has a largely local effect. In many cases, however, source control has both local-scale and river basin scale effects. An example here is the reduction in contamination levels in wetlands. The main benefits of source control can also become clear on a mainly river basin scale. This applies, for instance, when a reduction in toxic loads in fish, or a lower contaminant content of dredged material is at the objective.

Of course, the location of a management option along the river can restrict the impact it has for the river basin. For an option to have effects over the entire river basin, it is a necessary, but not sufficient, condition that it is not only implemented at downstream sites. For instance, riverbanks can be protected against erosion to reduce sedimentation in downstream areas (however, this type of measure often conflicts with nature development). The scale at which a management option is applied also determines how much of an effect it will have.

A site-specific measure for maintaining sufficient depth for navigation is the construction of groynes to alter hydro- and morphodynamics. Dredging in sedimentation areas is also a site-specific measure. If it is done properly, the dredging activity can have a beneficial outcome for ecology; optimal solutions for relocation, treatment, or storage of dredged material must be found.

To reduce the risks involved with dredged material disposal, the construction and use of confined disposal facilities (including disposal in sand or gravel extraction pits and capping them with a layer of clean material) is the most obvious measure. This can be regarded both as a site-specific measure (for small-volume facilities) or as a river basin measure (for large-volume facilities). Other site-specific measures are the various dredged material treatment or beneficial reuse options.

Long-term planning of structural development is a management option that operates on a larger scale (possibly, but not necessarily, a river basin scale). Realignment of dikes and reconstruction of the banks over large parts of the river are examples of this type of a measure.

For the interests of fish health and fishery activities, (site-specific) management options are capping of contaminated sites with clean sand or sediment to reduce contact between the water column and the contaminated sediment, or alteration of site-specific sediment dynamics to increase sedimentation and hence promote an enhanced natural capping of contaminated sites. As mentioned before, source control can have site-specific and river basin scale effects. Management options that are typical of a river basin approach include: the construction of fish ladders to allow fish to pass barrages, the implementation of measures that affect the current velocity at selected sites in the river basin to optimise living conditions for certain species, habitat compensation (when the cost of maintaining or restoring the presence of certain species in the area concerned is too high and a trade-off with other areas or other species is more efficient), the designation of protection zones in which fishing is not allowed (or is only allowed at restricted periods of time) and the initiation of negotiations with upstream stakeholders to take measures that upgrade the water quality downstream.

For flood prevention, a number of (site-specific and river basin) management options can be mentioned which aim at increasing the water discharge capacity of the river system, or increasing the water retention capacity of the catchment area. Examples are:

- lower flood plains (by excavation),
- create flowing side channels,
- reduce flow resistance of flood plains,
- lower the summer bed,
- implement realignment measures,
- reposition dikes,
- decrease the area of paved surfaces,
- restore forests,
- reconstruct riparian zones.

At river sections where other measures are insufficient or impossible, dikes can be raised to increase protection against flooding.

The measures that have been mentioned for mitigating flooding risks also have positive effects for safeguarding the interests of agriculture and the development of the area. Source control as a management option has been mentioned above.

Measures that can be taken to reduce the consequences of extreme drought periods, are the use (or construction) of barrages and sluices, use of alternative water resources for irrigation, use of alternative cooling methods in power plants and other industrial plants and the mobilisation of alternative power plants, or the importation of electricity.

For the maintenance or creation of a sustainable recreation level, measures can be taken that are beneficial to the quality development of water, sediment, soil and nature in the area. This includes measures such as capping contaminated sites, dredging, excavating, constructing sedimentation basins, source control, etc. Moreover, changes in land use (e.g. nature development) can be induced to stimulate recreation. At a river basin scale, management includes setting up or revising criteria for the relocation of (contaminated) sediment, land use, regulation of settlement of people and companies, etc.

A very important part of almost all management options discussed in this section is good communication with local or regional stakeholders (inhabitants, politicians, authorities, industries, etc.) on the risks that are present and the measures that have been, should be, or will be taken to reduce these risks; see also section 6. At a river basin scale, this communication has to take place at a higher political and management level than at the site-specific scale. An overview of the risk indicators and management options for the social and societal driving forces and risks that have been discussed is given in Table 3.

Table 3: Risk indicators and management options for the objective of maintaining economic viability

Example driving forces	Risk and impacts involved	Indicator of risk		Management option	
		Site-specific	River basin approach	Site-specific	River basin approach
Maintaining navigation and water transport (discharge) capacity	Economic losses from decreasing water depths, increased risk of flooding	Shallowing of channels / decrease in potential ship tonnage Decrease in harbour activities	Shallowing of channels / decrease in potential ship tonnage Decrease in harbour activities Use of alternative navigation routes	Alteration of sediment dynamics, groynes, river bank structure Dredging	Long-term planning of structural development
	Dredged material disposal	Local storage capacity Disposal costs Failure to comply with regulatory requirements	Decrease in maintenance dredging activities Volume limitations of disposal options Loss of resources (sediment as fertiliser, or construction material)	Confined disposal facilities Innovative treatment (cleaning, sand separation, immobilisation) and beneficial re-use	Confined disposal facilities Source control

Example driving forces	Risk and impacts involved	Indicator of risk		Management option	
		Site-specific	River basin approach	Site-specific	River basin approach
Sustaining fishery activities	Loss of economic function (fishery)	Decrease in fishery yield Toxic loads in fish Eutrophication Habitat change	Decrease in fishery yield Toxic loads in fish	Capping contaminated sites Alteration of site-specific sediment dynamics Environmental dredging Source control	Source control Fish ladders Affect current velocity Habitat compensation Designation of protection zones / fishing time restrictions Negotiations with upstream stakeholders
Flooding - catastrophic events or annually recurrent	Damages to infrastructure Loss of property (industrial and private) Increase in insurance costs	Increase in flood events Rise of average water level / river discharge Payment of insurance benefits	Increase in flood events Rise of average water level / river discharge Payment of insurance benefits	Increased discharge capacity by flood plain modification, side-channels, lowering summer bed, repositioning of dikes, realignment, etc. Increase in water retention capacity by reducing areas of paved surfaces, restoring forests, reconstruction of riparian zones Raising dikes	Increased discharge capacity by flood plain modification, side-channels, lowering summer bed, repositioning of dikes, realignment, etc. Increase in water retention capacity by reducing areas in paved surfaces, restoring forests, reconstruction of riparian zones Raising dikes
					Continued on next page

Example driving forces	Risk and impacts involved	Indicator of risk		Management option	
		Site-specific	River basin approach	Site-specific	River basin approach
Flooding - catastrophic and annual (continued)	Loss of economic function (agriculture)	Decrease in agriculture yield Toxic loads in crops, milk and meat	Decrease in agriculture yield Toxic loads in crops, milk and meat	See above (flood prevention) Source control	See above (flood prevention) Source control
Extreme drought	Reduced availability of cooling water (e.g. for power plants)	Decrease in agriculture yield Production loss of power plants Navigation problems	Decrease in agriculture yield Production loss of power plants Navigation problems	Use (or construct) barrages and sluices Alternative irrigation water resources Mobilise alternative power plants or import electricity Use alternative industrial cooling methods	Use (or construct) barrages and sluices Alternative irrigation water resources Mobilise alternative power plants or import electricity Use alternative industrial cooling methods
Recreation	Economic loss from decline in tourism	Number of visitors, overnight stays, restaurant turnover usage of yachting facilities (marinas) Eutrophication, bad smell Negative publicity, loss of clean environment awards	Number of visitors, overnight stays, restaurant turnover	Quality improvement measures Source control Local change in land use	Quality improvement measures Source control Changes in land use, settlement regulation Revision of relocation criteria

4. Sediment Management Objective: Ensuring Environmental Quality and Nature Development

4.1. Social and societal driving forces

The objective 'Ensuring environmental quality and nature development' is driven by at least three different social values or objectives which are partly reflected in the environmental regulations that have been put into force in the last 30 years: Maintaining ecosystem services, environmental ethics and the awareness of the socio-economic impact of environmental mismanagement.

The first of these social values is a human-centred attitude which intends to ensure the environment for public benefit, based on the recognition that humans are part of this environment, but acknowledges that their activities may lead to impairment of provisioning and supporting services that normally contribute to human well-being [21]. Provisioning services comprise food production which in the case of sediments applies to fisheries; supporting services of sediments are re-cycling of nutrients and degradation of organic matter.

The main contributors in alerting people's awareness that 'unabated pollution and unstabilised population are real threats to our way of life and to life itself' [22] were publications such as *Silent Spring* from Rachel Carson [23], 'Limits to Growth' [24], a report for the Club of Rome, and the 'Global 2000' report of the US Government [25]. The European water laws, which were mentioned in section 2, are all examples of regulations that aim at preventing human health risks.

Other environmental regulations, however, such as the Council Directive on the Conservation of Natural Habitats and of Wild Fauna and Flora, the Wild Birds Directive and the Environmental Impact Assessment Directive do not refer to the potential impact on humans, except perhaps in relation to their recreational function. Their main intention is the maintenance of environmental quality *per se* and they do not necessarily refer to related topics, to which most public concerns are directed. They are the result of the perception that the natural environment has a right to exist independent of human society so that its values ought to be respected and protected, regardless. What John Barkham described as 'environmental ethics' has been reflected since the 1970s in a number of multilateral environmental agreements [26]: The Convention on Wetlands of International Importance Especially as Waterfowl Habitat (Ramsar, 1971), the Convention Concerning the Protection of the World Cultural and Natural Heritage (World Heritage, 1972), the Convention on International Trade in Endangered Species of Wild Fauna and Flora (CITES, 1973), and the Convention on the Conservation of Migratory Species of Wild Animals (CMS,

1979). While the human population will not cease to exist if a number of species becomes extinct – we have survived the documented extinction of 27 species over the last 20 years (The World Conservation Unit 2004)[6] without any remarkable effect on our quality of life – biodiversity, endangered species and pristine habitats have become a value on their own, as has been claimed in the Principles of the World Charter for Nature (UN 1982).

This belief, however, is restrained by two attitudes:

a. People tend to give those issues that touch them emotionally a higher value. Preservation of whales has a much better chance of getting public support than the preservation of a widely unknown and ugly fish or invertebrate species. This needs to be addressed when the management objective of 'ensuring environmental quality and nature development' is discussed with regard to sediments. Public awareness of the ecological value of sediments is still low. Their perception is mainly restricted to its tendency to accumulate contaminants and being 'dirty', whereby the finer the sediments, the more they are perceived as being dirty, even though those sediments have the highest activities and ecological function in terms of e.g. degradation and remineralisation of organic material and biomass production in the environment. The unpleasant smell of anaerobic sediments due to hydrogen sulfide (H_2S) does not help to raise its value in people's minds. Despite this observation, the results of a cost-benefit analysis for the perception and valuation of clean sediments and biodiversity in the Netherlands indicated a willingness to pay for sediment remediation in return for positive effects on biodiversity [27].

b. Environmental ethics and the understanding of the interconnectivity of raising the standard of living and environmental quality are in most places secondary to the struggle for survival and against poverty. Long-term solutions that involve financial cuts are not acceptable to those who are struggling to feed themselves, and often not even for those who would only have to lower their standard of living in order to preserve their environment. Therefore, securing an acceptable living standard has to accompany activities for ensuring environmental quality, as has been stated in the Brundtland Report: "Sustainable development is development that <u>meets the needs of the present</u> without compromising the ability of future generations to meet their own needs" [28].

Accordingly, raising awareness and knowledge of the environmental role that sediments play, while at the same time aiming for sustainable solutions to ecological problems is an important step in sediment management.

After 'limiting negative impacts on humans' and 'environmental ethics', the third societal driving force is the financial impact of environmental

---

[6] http://www.iucn.org/themes/ssc/red_list_2004/English/newsrelease_EN.htm

mismanagement. The costs of global warming have been estimated to be higher than US $300,000 million annually due to more frequent tropical cyclones, loss of land due to rising sea levels and damage to fishing stocks, agriculture and water supplies [29]. Although the more direct driving costs to ensure environmental quality in Europe will be those that cover the efforts for compliance with the WFD, which have been discussed in section 2, indirect costs will also arise due to global warming impacts: In the Atlantic-dominated continental shelf, elevated temperatures will increase its seasonal amplitude, potentially leading to reduced discharges of rivers in summer and increased discharges in winter [30], with the respective effects for current velocities, morphodynamics, grain size distributions, dimensions and variations of estuarine conditions at the flow of the rivers and transport of suspended matter towards the sea, all of which will impact the biodiversity in rivers and the functioning of the nutrient cycles [31].

4.2. Risks involved

Risks perceived by humans in terms of environmental quality and nature development comprise the impairment of the two already mentioned ecosystem services: Fish production (provisioning service), re-cycling of nutrients and degradation of organic matter (supporting services).

Provisioning service:
The production of fish can be negatively influenced by the physical destruction of habitat, by changing abiotic parameters like pH, temperature, oxygen, current velocity, and by contaminating the environment, potentially causing acute or chronic effects.
Fish species depend on sediment during various life stages. Eels are bottom-dwelling, as are flatfish. Some fish species, such as salmon, lay their eggs on sediment and are hence susceptible to any deterioration of the sediment habitat (see below). Freshwater species threatened in Europe include the salmon (*Salmo salar*), the sturgeon (*Acipenser sp., Huso huso*), and the freshwater pearl mussel (*Margaritifera margaritifera*) [32].
Any impacts leading to a decline in fish abundance may not necessarily have affected the fish in the first place. The demersal fish species that are preferred in Northern Europe and North America, feed mainly on sediment-associated fauna. The FAO stated that "negative impacts on benthic communities may cause a decline in marine resources, including those exploited commercially" [33].
The smallest organisms in the benthic food chain are microbes that occur in total counts of between 10^9 and 10^{10} cells per cm^{-3} [34-36], compared to 10^5 to 10^6 cells per cm^3 in the water column [37]. Micro-organisms are fed upon by

multicellular organisms that are especially abundant at the upper sediment layer (see figure 5). Fenchel found that core samples from a transect perpendicular to the water's edge of a Danish beach yielded nearly 3,000 individuals of small metazoans which would pass through a sieve with a mesh size of 0.5 or 1 mm (Meiofauna) [38]. These organisms belonged to 71 species, of which 43 were nematodes. Other representatives in freshwater sediments can be ciliates, turbellarians and rotifers. A very rich interstitial fauna exists in the coastal ground water of sandy beaches. In silty and clayey sediments the character of the meiobenthos changes and nematodes, capable of burrowing, dominate. At the water-sediment interface, a rich fauna can be found with lots of different species, including juvenile specimens of macrofaunal species [38].

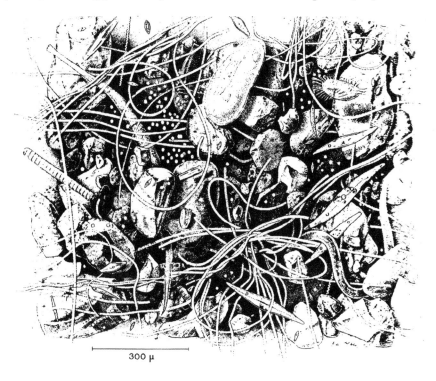

300 µ

Figure 5: Approximately 1 mm^2 of sediment surface with micro-organisms (from [39]

Examples of marine and/or freshwater macrofauna organisms are polychaetes, amphipods, insect larvae (freshwater), and echinoderms (marine). The differentiation between microfauna (1-100 µm) (bacteria, protophytes, protozoans), meiofauna (100 – 1000 µm) and macro or megafauna (>1000 µm) was suggested by Mare as an operational separation reflecting the sizes of sieves [40]. There may, however, be an ecological meaning to it, as the habitat that an organism can inhabit depends on its own size and – among other things – the grain size distribution of the sediment.

These habitats can be disturbed in different ways: In marine systems, most concern has been directed towards impacts of towed fishing gear like trawls and dredges on sediment [33]. Addition of sediment (e.g. by relocation or disposal of dredged material) or modification of the hydrodynamics of a river can lead to changes of grain size compositions.

The impact of quantities of sediment with a grain size below 2 mm on biota has been reviewed by Wood & Armitage [41]. They summarise different ways in which high concentrations of fine sediment can interfere with lotic fisheries: by clogging gill rakers and gill filaments [42]; reducing suitability of spawning habitat and hindering the development of fish eggs, larvae and juveniles [43]; modifying the natural migration patterns of fish [44]; and reducing the abundance of food available to fish due to an increase in turbidity (reduction of primary production, visibility) [42,45].

Benthic macroinvertebrates are assumed to be able to cope with changing quantities of suspended solids given the high variability of river waters. A risk, however, can arise from a permanent shift towards deposition of fine material as it accompanies agriculture and surface mining activities [41]. Many invertebrates rely on specific grain sizes for uptake and for habitat (e.g. burrowing organisms). Alteration of substrate composition can change its suitability for both [46,47]. In addition, fine sediments can block respiratory organs [48] and impede filter feeding [49]. There has been little research on the effect on single taxa [41]. Suspended fine sediments additionally impact primary production as it decreases the light penetration in the water. It has also been shown to damage macrophyte leaves and stems due to abrasion [50].

A decrease in abundance of species can also arise from contaminants that effect organisms in the food web or the fish themselves. Examples of such contaminants are manifold: Substances like PCBs, chlorobenzenes, dioxins, that are persistent and bioaccumulating, leading to harmful concentrations in top-predators like penguins, dolphins, whales, polar bears, etc.

A monitoring programme conducted in the Elbe River one year after the flood in 2002, detected hexachlorocyclohexane (HCH) concentrations in dermersal breams which exceeded the maximal threshold value of 10 ng/g of β-HCH 18 times [51]. The contaminants originated from the area of former lindane (γ-HCH) production in Bitterfeld in the former GDR. The flood probably eroded contaminated sediment and soil and transported it via the tributaries to the Elbe River. Chemical analysis of sediment samples over a time span of 10 years clearly showed that the tributary Mulde at Bitterfeld is the main source of HCH contamination in the Elbe River [52]. 60,000 t of industrial waste rich in HCH-derivatives are still stored in a pit in this region [53], demonstrating that even though the concentration in the breams was probably not threatening the fish population in the Elbe, exposure of demersal fish to old industrial legacies especially during floods continues to be an issue.

To determine the risk posed by contaminated sediments is complicated due to the "cocktail" of various compounds that accumulate in organisms over time and which may show synergistic or additive effects. Different substances are most likely to be in different stages of ageing and residual formation, hence with different bioavailability for organisms [54,55]. Micro-organisms have been shown to actively increase bioavailability of bound chemicals by different methods, such as adhesion to substrate sources, secretion of surfactants and change in the affinity of their uptake systems [56]. The uncertainty of the risk for river biota increases because of chemicals of unknown toxicological effects: The European Inventory of Existing Commercial Chemical Substances (EINECS) lists 100,000 chemicals, 75% of which have not been toxicologically tested [57]. Previously unexpected or unknown effects of environmental contaminants further increase the uncertainty, a recent example being that of the endocrine-disrupting substances in the aquatic environment that are linked with sexual disruption in aquatic animals and considered an emerging issue of concern [58]. A number of studies have now been carried out on freshwater and estuarine systems in Europe and endocrine disruption has been noted in fish exposed to effluent from sewage treatment plants. The main observation is the feminisation of males, including the induction of vitellogenin (an egg yolk protein) and abnormal gonadal development. The effects on populations are, at present, unclear but it is generally considered to be mainly due to natural and synthetic oestrogens from domestic sewage. The most undisputed evidence for endocrine-disrupting chemicals effecting wildlife populations is that for organo-tin compounds. Organo-tin compounds were first used in anti-fouling paints in the 1960s and have now been shown to cause imposex (penis formation induced in females) in over 100 species of marine molluscs [58].

According to the WWF report *Water and Wetland Index* in which 16 European Countries[7] were assessed [32], 50 out of 69 river stretches across Europe are still rated as poor in terms of ecological quality. The trends for threatened species are only positive in Denmark. Red Lists of threatened species are inadequately updated and the national information on current threats on biodiversity is very poor.

---

[7] Austria, Belgium (Flanders + Wallonia), Bulgaria, Denmark, Estonia, Finland, France, Germany, Greece, Hungary, Slovakia, Spain, Sweden, Switzerland, Turkey, UK (England, Wales, Scotland, Northern Ireland)

Without attempting to detract from these still existing issues of concern, the situation of many rivers in Europe has significantly improved in most countries [58], especially compared to the situation 20–30 years ago. In the Rhine catchment area for example, 1969 and the early 1970s were the time periods of highest industrial and communal wastewater load to the Rhine, corresponding to a minimum of biodiversity in macrozoobenthos. Long stretches of the Rhine were devoid of all animals and on even larger stretches , no insects, turbellids and crustacea, the most sensitive families, could be found. Extensive restoration measures, especially the start-up of communal and industrial wastewater treatment plants along the Rhine and its tributaries, led to a considerable recolonisation of this river [59].

Supporting Services: the breakdown of particulate organic matter, the uptake of dissolved organic carbon and the (re-)cycling of inorganic nutrients.
Sediments are essentially a heterotrophic system, in which particulate organic matter that has been produced in the water phase (e.g. settled phytoplankton, zooplankton) or derived from land (wood, leaves, but also waste, etc.) is broken down by a cooperation of benthic communities. As early as 1971, Kaushik & Hynes found that the first steps of leave decomposition are mainly done by fungi rather than by bacteria and that leaf-shredding invertebrates prefer leaves colonised by fungi, which present an additional food source [60].
Invertebrates process the coarse material into fine grains which is then further degraded – at a certain stage by fungi and bacteria.
The most important electron acceptor during the degradation of organic matter is oxygen. Due to its low solubility in water, it's diffusion into sediment is limited. High activity of aerobic, oxygen-respiring micro-organisms, may quickly diminish the surrounding oxygen concentration. If this process occurs faster than oxygen can be supplied from the water phase above, anoxic conditions establish. The remaining organic substances will be metabolised using other electron acceptors like NO_3^-, MnO_2, $Fe(OH)_3$, S, SO_4^{2-}. If these are not available, methanogenesis occurs.
These complex environmental processes are based on a cooperative consortium of organisms with different functions and different capabilities. A current discussion suggests to make use of potentially impacted environmental functions as an indicator of environmental quality. A study to identify the impact, that loss of environmental function can have on an ecosystem was conducted by Wallace et al. [61]. Of two streams with many species of leaf-shredding invertebrates, including insects, that transformed coarse leaf-litter into smaller particles, the population of stream-dwelling insects in one stream was strongly reduced by low-dose application of an insecticide. Compared to the second stream, serving as a reference, shredder secondary production in the

test-stream was lowered to 25 % by this treatment. Standing stocks of leaf litter increased and organic carbon export from this watershed decreased dramatically, probably lowering animal production in downstream food webs [62].

While fungi seem to be more important in the break down of leaf litter, bacteria play a main role in the mineralisation of dissolved organic carbon (DOC). A main step of the benthic microbial loop consists of the uptake of non-refractory DOC also produced by the enzymatic hydrolysis of organic particles which is mostly unavailable to higher organisms. It is partly oxidised to CO_2 and partly used to build up organic matter (microbial biomass) which is then provided to higher food web levels. This important "recycling" of dissolved organic matter in streams and marine waters has become famous as the "microbial loop" [63,64]. The crucial function of the benthic microbial loop in the functioning of shallow coastal ecosystems has been described by Manini et al. [65].

Carbon cycling involves activities of both micro- and macro-organisms. Other key nutrients, however, such as nitrogen, sulphur and iron are cycled exclusively by micro-organisms and all rely on some transformations on the presence of anoxic environments, such as anaerobic sediments.

In the case of nitrogenous compounds, ammonia (NH_3) is produced by the breakdown of nitrogenous organic matter such as amino acids and nucleic acids. It still has the same reduced state as nitrogen organic matter and is therefore the most energy-sufficient nitrogen source of biomass production for plants. High, toxic concentrations, however, can be reached in sediments, as ammonium is relatively stable under anoxic conditions and it is in this form, that nitrogen predominates in most sediments.

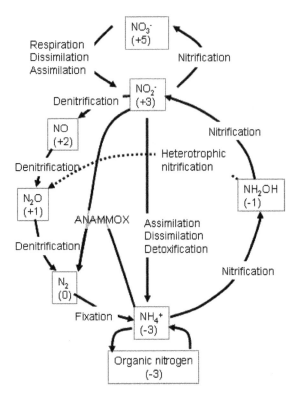

Figure 6: The biological nitrogen cycle. Figures in parentheses denote the nitrogen oxidation state ANAMMOX: anaerobic oxidation of ammonia with nitrite (redrawn from [66], modified).

Under oxic conditions, e.g. in the water column or in the oxidised upper sediment layer, chemolithotrophic bacteria, that build organic matter from CO_2, use ammonia and nitrite (NO_2^-) as electron donors in a process called nitrification, The complete oxidation from ammonia to nitrate (NO_3^-) is carried out by two different groups of "nitrifying bacteria" that act in concert: Nitrsomonas and Nitrobacter. Nitrogen compounds are removed from the system by denitrification, which occurs under anaerobic conditions: Nitrate acts as an alternative electron acceptor during anaerobic respiration. Nitrate is thereby reduced to gaseous compounds N_2 or N_2O. This process is the main process by which gaseous N_2 is formed biologically.

Thus nutrient cycling requires a large number of different organisms from different functional groups. It is a prime example of "functional diversity" in action. Conversely, dysfunctions in nutrient cycling, leading to for example, eutrophication, can have severe negative effects on biodiversity [67]. Strong interference with the nitrogen cycle in this respect has been ongoing for years. From 1860 to the early 1990s, anthropogenic creation of reactive nitrogen compounds like NO_x and NH_3 increased globally from 15 Tg N/yr to 156 Tg N/yr due to two anthropogenic activities: Food production, producing reactive nitrogen species as plant nutrients on purpose, and energy production, creating it undeliberately during combustion of fossil fuels. [68]. Even though approximately 78% of the Earth's atmosphere is diatomic nitrogen (N_2), this is unavailable to most organisms because of the strength of the triple bond that holds the two nitrogen atoms together. Only a limited number of bacteria and archae are able to convert diatomic nitrogen into reactive species in a process called nitrogen fixation. With various reactive nitrogen compounds, numerous mechanisms for interspecies conversion, and a variety of environmental transport/storage processes, nitrogen has arguably the most complex cycle of all the major elements. This complexity challenges the tracking of anthropogenic nitrogen through environmental reservoirs. Nevertheless, this work is necessary because of nitrogen's role in all living systems and in several environmental issues (e.g., greenhouse effect, smog, stratospheric ozone depletion, acid deposition, coastal eutrophication and productivity of freshwaters, marine waters, and terrestrial ecosystems) [68]. Up until now, emissions of reactive species in the atmosphere have increased much more rapidly than their riverine discharges to the coastal zone. The relatively limited response of riverine systems to increases of atmospheric reactive nitrogen are most probably caused by the ability of terrestrial ecosystems to accumulate reactive N compounds, and the fact that significant amounts of nitrogen added to terrestrial systems are denitrified either within the system or in the stream/river continuum prior to transport to the coast. Rough assumptions estimate that 30 to 70 % of the nitrogen that enters a river is denitrified in the sediment or at suspended particles and that about 50 % of the remaining is denitrified in the estuary. However, Galloway et al. expect the extend to which nitrogen enters riverine systems to increase as terrestrial sinks become increasingly saturated and the continued removal of wetland and riparian landscapes reduces denitrification [68].

4.3. Indicators of risk

4.3.1. Site-specific
Site-specific risk indicators are measured at a specific location, e.g. in a specific sediment sample, and provide information on the properties at that site. Such properties can be contaminant load, change in biodiversity, ecotoxicological effects or positive responses in biomarkers. From these properties, the state of the environment has to be evaluated and the degree of its risk assessed. As has

been pointed out by Tannenbaum, most so-called risk assessments describe an impact instead of risk [69], meaning that the situation at a contaminated soil site, for example, no longer poses a threat to the surrounding organisms because those that persist have adapted and more sensitive ones have disappeared. In a stable environment without changes in exposure and without hazardous effects on the biota, no risk needs to be calculated, only an impact can be described. The situation for sediments, however, is different from soil because the moving water column above continuously exposes new organisms to bioavailable contaminants in the sediment. However, bioavailability as well as exposure change because of ageing processes [54,55,70], and hence the risk and its variation with time to previously unexposed organisms needs to be assessed. Additionally, management activities such as capping or dredging can have a considerable effect on a (contaminated) sediment. Therefore, decision making depends on an assessment of the potential impacts, and hence the risk, that these actions may have on the ecosystem. Given either a new dumping site, which may affect the surrounding environment, or the risk for the aqueous environment on top of the sediment, parameters that could indicate an environmental risk should be of a chemical, ecotoxicological or ecological nature (see *Chapter 5 - Risk Assessment Approaches in European Countries*)

Concentrations of contaminants in sediments have long been used as the only quality criterion and in some countries they still are. However, the large number of substances produced by industry, the financial expenditure that would be necessary to cover the chemical analyses of this wide range of potentially occurring contaminants, and the realisation that the question of concern should be whether or not significant risks (potential or actual effects on ecology or human health) exist, not whether a chemical guideline is exceeded, have led to a discussion on new approaches to assessing risk in sediments [71]. Combining multiple lines of evidence (LoE's) in order to increase the accuracy of the assessment is gaining interest[8]. Lines of evidence comprise environmental descriptors such as contaminant load and geochemical characteristics (grain size, oxygen concentration, etc.), as well as ecotoxicological responses and data on the (benthic) community which could comprise, for example, biodiversity, body burdens of contaminants, changes in behaviour and symptoms of diseases. As each of these LoE's has its flaws, uncertainty can be reduced by applying a *weight of evidence approach*, as has been suggested by Burton [72]. Multiple lines of evidence are linked appropriately (environmental descriptors vs. biological responses) in a more quantitative fashion, and response patterns are used to assess the risk of sediments in order to increase the probability that the

---

[8] see HERA, Vol. 8 (7) December 2002, Debate and Commentary section, 10 contributions by various authors

real environmental situation is described and a more certain risk prognosis can be given.

Loss of non-migrating species or the occurrence of lesions or necrosis also indicates a risk that exists at a specific site. This, however, does not necessarily point to a local source disturbance, because it can well be upstream. Losses of sensitive species are often accompanied by an invasion of other, more resistant organisms that take over the empty ecological niche. This leads to changes in biodiversity and the abundance of species. This should be monitored, although interpretation is often impeded by the lack of knowledge on natural variability.

The degree of eutrophication and biogeochemical properties of the sediment such as chemical and biological oxygen demand, or a change in redox conditions can be seen as local indicators for disturbances in ecological functioning for example by interference with the nutrient cycles.

Changes in sediment dynamics and hydrodynamics may have a direct effect on benthos communities, which may be caused by changes in the river's morphology or discharges. The consequences can increase sediment loading (suspended or depositing sediment), alter and remove habitat including stream bank and riparian vegetation, disrupt fish passage and release deleterious substances into the water because of resuspension of contaminated material. Monitoring sediment dynamics and hydrodynamics at a site could therefore be used to check for physical changes in the environment that may affect the local benthic community and thereby disrupt the benthopelagic cycle.

4.3.2. River basin scale

Ecological changes that endanger a whole river system have some of the same indicators as the site-specific risks, differing in that they spread over a much larger area. For example, a temperature elevation at a local site may be caused by a factory discharge. If the same temperature elevation, however, can be measured along the river basin, species that are temperature-tolerant will have an advantage over those that depend on a limited temperature range and may be forced to move to the new gradient. Substitutions of species up- or downstream, or the loss of species that come from the sea and wander upstream, the most famous example of which is the salmon, give a good indication that quality is changing.

4.4. Management options

4.4.1. Site-specific

In order to follow the management objective of ensuring environmental quality, the first option considered is source control. If a risk has been identified at a specific site, but no source could be detected, site-specific management must turn to river basin management, as those problems can only be solved on a catchment level, e.g. by building water treatment facilities.

If local activities are responsible, changes in industrial processes at local plants, reduction of emissions from communal sources and agricultural practices (nutrient and pesticide emissions) and, if necessary, subsidised changes in land use could be options to reduce local sources of environmental deterioration.

If the contaminated sediment poses a risk that needs to be taken care of, decisions should be made on whether its degree of contamination can be reduced, e.g. by in-situ treatment, or whether the exposure pathways need to be disconnected. Successes of in-situ treatment methods in Europe are scarce, as they are normally limited to specific contaminants and can seldom be applied to a mixture of different substances. Adsorptive barriers and deployment of capping structures tend to reduce the exposure of organisms and at the same time avoid environmental dredging or excavation activities, which often lead to a large amount of resuspension of material. While environmental dredging is conducted frequently in the United States, it is seldom done in Europe. In any case, habitat for benthic or demersal organisms is destroyed directly by substrate removal and indirectly when the sediment eventually settles again on the stream bottom, where it can bury important food, spawning, and cover habitat. Habitat compensation is one management option that needs to be considered when natural sediment has to be removed. In some countries, habitat compensation is frequently required for any management activity that destroys environmentally valuable habitats and forbids these activities if compensation is not possible (e.g. §19 BNatSchG – Federal regulation for Nature Protection).

If no immediate action is necessary, *monitored natural recovery* would be the management option of choice. Monitored natural recovery – sometimes described as 'natural attenuation' or as 'do-nothing' option – is based on the observation that bioavailability of adsorbed contaminants decreases with time. This needs to be verified by regular site-specific monitoring, so that interactive effects can be detected and impacts that may be caused by environmental changes can be predicted, such as increasing current velocities. Sometimes, natural recovery can be stimulated by creating the right environment conditions (enhanced natural attenuation) .

4.4.2. River basin scale
The first priority of management options on a river basin scale is to identify those sources with the highest ecological risk for the river ecosystem and eliminate those along the river basin (see chapter 3 - *Strategic Frameworks for Managing Sediment Risk at Basin and Site-Specific Scale*). Identification of the basin-specific substances of concern, their biological effects and interactions is

a management option on the river basin scale. For source control, enforcing the Polluter Pays Principle, as already recommended in the WFD and to be implemented into European regulation in 2009, will support reduction of future emissions. Table 4 gives an overview of the risk indicators and management options for ensuring environmental quality and nature development.

Table 4: Risk indicators and management options for the objective of ensuring environmental quality and nature development

Example driving forces	Risk and impacts involved	Indicator of risk		Management option	
		Site-specific	River basin approach	Site-specific	River basin approach
maintaining ecosystem services Environmental ethics Awareness of costs of mis-management	Harming human health Extinction of species Destruction of habitat Reduction of ecological function Degradation of water quality Conflict with other objectives	Contaminant load Ecotoxicological effects / biomarkers Change in biodiversity / abundance Loss of species, invading species Increase in fish lesions / necrosis Change in sediment dynamics / Hydrodynamics Eutrophication Biogeophysical parameter characteristics	Change in migrating fish species (e.g. salmon) / loss of indicator species Change in biodiversity / abundance Physicochemical parameters (O_2, H_2S, etc) Imbalance of nutrient cycles (Eutrophication) Sediment / contaminant budget	Source control Turn to RBM Subsidised change in land use Exposure reduction: Adsorptive barriers Excavation / dredging Capping In-situ treatment Natural attenuation Habitat compensation	Source control, reduction of diffuse pollution, Trend monitoring Revision of industrial and agricultural policies Identification of responsible contaminants Enforcement of Polluter Pays Principle, or injured party pays to accelerate solutions Enhancement of transboundary co-operation, implementation of basin-wide management plans

5. Sediment Management Objective: Securing Quality of Human Life

5.1. Social and societal driving forces

The specific management objective of securing the quality of human life is driven by the need to care for the individual as well as the social system's well-being. The individual's well-being comprises good health, security of income, jobs and education, as well as social justice and life satisfaction. The social system's well-being, according Miranda *et al.*, includes sustainability (e.g. provisions for future generations), social justice (e.g. income distribution, access to environmental quality), and absence of war, famine, and social unrest [73]. The threat of catastrophes – either natural or man-made – however, affect both the individual's well-being and the society's, also because it raises questions about responsibilities and confidence in (governmental) institutions. Catastrophes of this kind include flooding, but also industrial accidents.

According to the international disaster database (EM-DAT), floods comprised 43% of all disaster events for the period 1998–2002. During this period, Europe suffered approximately 100 major floods, causing some 700 fatalities, the displacement of roughly half a million people and at least 25 billion euros in insured economic losses [74]. The risk of flooding results from a combination of natural factors and human interference. Human actions can influence flooding by increasing the possible impact of high discharges (e.g. by occupation of flood plains) and/or by affecting the run-off patterns [75]. Hydraulic engineering works such as river canalisation may lessen the flood impact in one area but worsen its effect in another. Canalisation generally changes a heterogeneous meandering river into a homogenous straight channel with an increased bed slope, uniform flow conditions and less habitat diversity compared to the undisturbed situation [75].

River regulation has been undertaken to the largest extent in western and southern Europe (see Box 1: the adaptation of the Rhine to human needs and its consequences). In countries such as Belgium, England and Wales, and Denmark, the percentage of river reaches that are still in a natural state are between 0 and 20%. In contrast, in countries such as Poland, Estonia and Norway, 70–100% of the reaches of many rivers have remained in a natural state [76]. However, in almost all the European countries, the natural balance between rivers and their flood plains that hold back water during flood periods has been disturbed [76]. Catastrophic flood events are a driving force that falls under more than one management objective, as they also have economic and ecological consequences, respectively, but they certainly have impacts on public safety, linking this driving force to the management objective of securing the quality of human life.

Box 1: Adaptation of the Rhine to human needs and its consequences

Until the beginning of the 19th century, the Upper Rhine still had the characteristics of a natural river. The first and most momentous adaptation to human needs started in 1817 with the realisation of the "rectification plans" of Johann Gottlieb Tulla. Tulla wanted to reduce the destroying effects of the Rhine waters by straightening the river, by demolition of dams in the water, and by cutting off its small side arms [77]. He aimed for an increase of cultivable land, improvement of soil and geographical fixation of the border to France [78]. Due to these measures, the former width of the Rhine of up to 12 km was narrowed down to 200 to 300 m. Because the bank erosion was impeded by embankments and groynes, bed erosion increased. In 150 years, the depth of the Rhine increased by 7 m [79].

In the years that followed Tullas activities, the agricultural use of former flood plains was intensified through drainage measures, and 12 barrages were built along the Upper Rhine in order to use hydropower and facilitate shipping. In the area of the barrages, the river lost 60% of its original flood plains. In addition, they reduced the natural sediment transport to a large extent. Since 1977, material is artificially added below the last barrage Iffezheim in order to reduce bed erosion [80].

The Lower Rhine has also increased its depth due to hydraulic engineering measures during the last 100 years. The sediment dynamics in this area were further disturbed by extensive gravel withdrawal from the river and by subsidence of the riverbed as a consequence of underground mining of salt and coal. In the forming depressions, sediment accumulates and is unavailable for the river section downstream [79].

An unexpected effect of the straightening of the river was that the time a flood wave needed to travel down the Upper Rhine was reduced and now overlaps with the arrival of flood waves from the tributaries at the respective confluences. This elevates the peak of the water-level of the Rhine floods and increases their impacts below the confluence of the Neckar [80].

Recently, measures are planned to redo these early modifications of the Rhine river by re-creating flood plains and re-locating dams. In the area of the Rhine-bordering Federal State Nordrhein-Westfalen alone, the provision of potentially flooded areas with a retention volume of 170 Mio. m^3 were planned in 1996 [81].

5.2. Risks involved

Risks and impacts of sediment issues on the quality of human life comprise quality issues, quantity issues and the perception of risk. An experienced or continuously perceived risk can ruin the individual's well-being for a long time. Floods, for example, cause immediate drowning, injuries, and financial losses, but, probably most important with floods in Europe, they cause an increased incidence of common mental health disorders. Anxiety and depression may last for months and possibly even years after the flood event, often not being recognised in its connection to the natural disaster [82].

Even in cases where from a scientific standpoint risk has disappeared (or never existed in the first place), perceived risks are real to people and need to be addressed in management decisions. When a bridge over the Israeli river Kishon was destroyed and several people died after they fell into the water, media claimed that they were poisoned by the contact with sediment rather than drowned, drawing a great deal of attention to the issue of sediment contamination (Danny Sherban, Yodfat Engineers Ltd, Israel, personal communication). Even though it is relatively certain that sediment contamination was not the cause of the deaths, the high level of attention and concern that this incident raised had to be dealt with.

If risks are not perceived as threatening, but a situation is experienced as inconvenient, this also has an impact on public welfare and it has economic consequences. The knowledge that an area is contaminated keeps people from enjoying its recreational value. If the effects of the contamination can be clearly appreciated (a film of oil on the water surface, a strange odour), this has a high negative impact on aesthetics (scenic beauty and public appreciation), even though no direct threat to human health is involved.

It is, however, difficult to establish causal relationships between human health and environmental contamination. In many cases, contamination is just one of several factors including diet and other lifestyle choices, not to mention genetics, which influence whether an exposed person will ever become sick [83]. With regard to sediments, it is even more challenging when considering the different potential exposure pathways and contaminant cocktails. Nonetheless, human exposure to contaminated sediments is a central part of tiered approaches to risk assessment and risk management when evaluating sediment management options. As presented in Chapter 5 (*Risk Assessment Approaches in European Countries*), human exposure to contaminated sediments may occur through direct exposure via sediment ingestion, surface water ingestion, ingestion of suspended matter, dermal uptake via sediments and surface water, and indirect exposure via drinking water that has been affected by contaminated sediment and via consumption of contaminated food.

In general, direct exposure, even to highly contaminated sediments, is usually too short to create a risk for human health. During swimming, the average amount that people take in orally is relatively low. Nevertheless, this risk, especially for small children playing in mud or muddy waters, should not be excluded. The concerns of Greenpeace on the occasion of the Elbe-Badetag (Elbe-bathing day) in 2002 ('Im Gift schwimmen' (engl.: *'swimming in poison'*); Greenpeace magazine news Hamburg, May 2002) – although portrayed very sensationally – were shared by scientists and some regulators. In most areas, however, concern with regard to adverse effects of recreational water activities is related to risks caused by pathogenic bacteria (e.g. the swimming-related fact sheet of the Maryland Department of Health & Mental Hygiene).

In Europe, accumulation of toxic substances in animals that live in direct contact with sediments and the biomagnification through the food chain is the most significant exposure route for contaminants in sediment. Until today, the consumption of fish from many European rivers is still not recommended and restricted. This especially concerns fish that live in direct contact with sediment and have a high fat content such as eels. The PCB levels (PCB 153 and PCB 138) in eels from the Mosel, a French-German river, exceeded the allowed maximum of 0.3 mg/kg and accordingly the environment agency recommended not to consume this fish species [84].

Another indirect exposure pathway is the intake of drinking water that has been contaminated after having been in contact with contaminated sediment and water. Water for public supply comes from two primary sources: surface water or groundwater. The relative portion varies depending on the natural conditions and the characteristics of water uses in each country (Table 5). In countries with extensive groundwater reservoirs (e.g. Iceland, Austria), a major part of total abstractions comes from this source, compared with less than 15% in the Netherlands and Finland [85]. While surface water often shows elevated concentrations of unwanted substances, groundwater is of naturally good quality and very little or no treatment is needed to make it suitable for drinking. In case of overexploitation of groundwater resources, however, the groundwater level can be drawn down, which can influence the movement of water with a change in quality within an aquifer. Substantial water level lowering can cause significant quality changes, e.g. pollution because of potential increased exposure of previously unpolluted groundwater to polluted groundwater (typically in shallow layers) [86]. This is of particular concern in the Netherlands, where the high ground water table is in close contact with contaminated sediments.

Table 5: Apportionment of public water supply between groundwaters and surface waters [85].

Country	Groundwater (Percentage)	Surface water (Percentage)
Austria	30%	70%
Bulgaria	7%	93%
Switzerland	34%	66%
Cyprus	67%	33%
Czech Republic	28%	72%
Germany	16%	84%
Denmark	97%	3%
Estonia	17%	83%
Spain	13%	87%
Finland	12%	88%
France	19%	81%
Greece (est.)	36%	64%
Hungary	3%	97%
Ireland	12%	88%
Iceland	97%	3%
Lithuania	5%	95%
Luxembourg	52%	48%
Latvia	43%	57%
Macedonia	1%	99%
Malta	100%	0%
Netherlands	11%	89%
Norway	20%	80%
Poland	22%	78%
Portugal (est.)	86%	14%
Romania	12%	88%
Sweden	23%	77%
Slovenia	15%	85%
Slovakia	40%	60%
Turkey (est.)	15%	85%
England and Wales (NUTS95)	15%	85%

Legend: est. = estimated

Apart from the quality issue, sediment quantity can negatively affect quality of life by changing the geomorphology of recreational sites such as beaches.

When sediment accumulates in harbour basins, where it has to be dredged again in order to secure navigation, this can lead to a substantial economic burden on cities with large ports such as Rotterdam or Hamburg. In Hamburg, the expense that is incurred by maintenance dredging in the harbour and in the Hamburg-district of the Elbe amounts to a total of 18–20 million euros per year. On the other side, erosion of sediment around bridge girders or piers can also lead to substantial risks, as has been experienced when the Entre-os-Rios bridge that partially collapsed in Portugal in 2001 (see Box. 2: The collapse of the bridge Entre-os-Rios across the Douro River).

Box 2: The collapse of the bridge Entre-os-Rios across the Douro River
(by Luís Ivens Portela, Research Officer, National Laboratory of Civil Engineering (LNEC) Portugal)
On March 4, 2001, the bridge at Entre-os-Rios across the Douro River suffered a partial collapse, causing a bus and several vehicles to fall into the river. About 60 people were killed. The mechanism for the failure of the bridge, built in 1885, was the fall of one of the piers due to scour. An enquiry revealed that over the last three decades the longitudinal profile of the riverbed experienced a strong and generalised lowering, about 15 m at the bridge site, but reaching 28 m at other sites. This evolution was certainly due to the combination of two main factors: the activity of aggregate extraction from the riverbed and the reduction of sediment discharge caused by sediment retention in the reservoirs upstream. At the time of the accident, the severe river flow conditions (discharge 8,000 m^3/s) probably caused an additional lowering of the riverbed, albeit temporary. The enquiry found that the activities of aggregate extraction were conducted without the support of plans and technical studies required by existing legislation. There was also an apparent lack of adequate and effective supervision of these activities. One of the main recommendations of the enquiry was the need to define and enforce an integrated approach to sediment management in the Douro River.

5.3. Indicators of risk

The risk indicators for flooding events (in themselves) include an increase in the number of flooding events, a rise in the average water level, the payment benefits as well as increased insurance premiums, as listed in table 3. With regard to the risk to public safety as a result of flood events, these examples of risk indicators can be expanded also to the number of fatalities, the relocation or mobility of inhabitants as well as negative public response in opinion surveys. These risk indicators could apply to both the site-specific and the river basin approach. However, a distinction can be made for risk indicators that only apply to the risk of public safety at the river basin approach. The river basin approach accumulates more data and places events in a statistical perspective. Therefore, it is possible to have additional risk indicators that analyse trends such as claims for compensation from the governments or the drop in property values.

Historically, human exposure has often been estimated through environmental measurements of ambient pollutant concentrations. Therefore, monitoring ambient pollutant levels is critical to measuring exposure for several pollutants [83]. This approach, combined with information on acceptable or tolerable daily intake of contaminants, has been used to establish water and sediment quality standards that often become incorporated into regulatory legislation. Exceedance of these threshold values is considered an indicator of risk for human health.

Several environmental pollutants can accumulate in the body over time, often with increasing risk or harm. These pollutants and their metabolites leave residues in the body that can be measured, usually in the blood or urine. These residues reflect the amount of the pollutant in the environment that actually gets into the body. The approach of measuring pollutant levels in tissue or fluid samples from organisms or individual people is called biomonitoring. Environmental exposure to mercury, for example, is of concern as it can be transformed into methylmercury by bacteria in sediments and then move up the food chain, accumulating in fish, which are a major source of exposure for people. Methylmercury has also been associated with harmful effects on the nervous system, especially in a developing foetus [83].

Another indicator of risk to human health includes the number of fish advisories in place. National or local authorities issue fish consumption advisories and safe eating guidelines for waters in order to inform people of the recommended level of consumption of fish caught in local waters.

Actual incidences of food poisoning or infections are additional examples of risk indicators at the site-specific scale, whereas statistical trends of food poisoning or infections are examples of risk indicators at the river basin scale.

Although recreation refers to the recreational activities of the inhabitants of an area and not tourists that use an area, the damage to recreational activities can be assessed using some of the indicators that are applicable to the economic loss from a decline in tourism, as presented in section 3: usage of marinas and other recreational facilities or the loss of blue flag status for beaches. Indicators of risk that are appropriate for the more aesthetic aspects related to public welfare or well-being include the occurrence of eutrophication, noxious odours or negative press coverage.

5.4. *Management options*

In order to secure the quality of human life in the event of a flood, public safety must be ensured. Most of the management options available to do this have already been presented in section 3. For example, flood protection can be accomplished by storing flood waters in specially constructed storage or detention reservoirs or by modifying the river to accommodate the flood flows within the bank. This can be achieved by widening or dredging the river and by the construction of flood banks or levees (quays) adjacent to the river.

An additional flood protection management option that is closely linked to the quality of human life is ensuring public education, communication, and involvement in the decision-making process. These rights are laid out in The Aarhus Convention, giving stakeholders the right to obtain information on the environment, the right to justice in environmental matters and the right to participate in decisions that affect the government [87]. The right covers decisions on whether to allow specific activities (for example the construction of dams), plans and programmes that affect the environment, as well as policies and laws. Stakeholder participation can lead to decisions that better reflect the needs of the people and that last longer, as well as decisions that have greater validity (www.unece.org/env/pp). (see section 6).

Decisions on management options to ensure human health are based on the information available regarding ambient pollutant concentrations. This information is obtained through monitoring programs that include measuring concentrations of contaminants and pathogens in water, sediment and biota (fish). Biomonitoring that is initiated when there is an indication of risk can itself be a management option – to indicate sources and monitor improvement with time.

If pollutant levels in fish and shellfish are unacceptable, issuing advisories can be an effective management option. Bans can be issued prohibiting all fishing, or advisories can be issued concerning which species and sizes of fish to avoid. By discouraging fishing as well as the consumption of fish once caught, public information is being provided.

Elevated concentrations of contaminants in agricultural products as a result of crops grown, or cattle grazing on flood plains that have received contaminated sediment after a flood event are also of concern. Management options include issuing advisories, increasing efforts to change agricultural land use at the river basin scale and banning certain agricultural products.

Safeguarding drinking water is essential. Therefore, if elevated concentrations of contaminants or bacteria in either groundwater drinking water wells or surface water drinking water reservoirs are measured, management choices must immediately be implemented. These include dissemination of information, the use of other sources of water for drinking water and the implementation of additional drinking water treatment processes. Ultimately, the most desirable management option is source control. Other options are technical remediation solutions or the installation of adsorptive barriers.

Lastly, risk communication is an important management tool that should always be implemented together with other management solutions. Risk perception and communication is presented in detail in section 6. An overview is given in Table 6.

Table 6: Risk indicators and management options for the objective of securing quality of human life

Example driving forces	Risk and impacts involved	Indicator of risk		Management option	
		Site-specific	River basin approach	Site-specific	River basin approach
Flood events, caused by change in land use, global climate changes	Endangered human safety Economical impacts High river discharge, flooding, sedimentation of contaminated material in flood plains;	Negative public responses in opinion surveys Payment of insurance benefits after floods Mobility of people due to perception of insecurity	Claims for compensation from the governments Collapse in property values Payment of insurance benefits Cost of incidents for society	Public communication and involvement Flood plain modification Realignment Alteration of storage / discharge capacity Raising dikes River deepening	Public communication / education and involvement Reduction of extent of paved surfaces Flood plain modification Early warning systems Restrictions of land use River deepening
Public welfare (*)	Harm to human health Recreation (**)	Incidences (e.g. food poisoning, infections) Evidence or suspicion for bio-magnification Use restrictions	Evidence or suspicion for bio-magnification Statistical trends of food poisoning, infections	Biomagnification studies Monitoring of pathogens Water and sediment quality control Recreation restrictions	Source control Legal restrictions to use and consumption of fish and agricultural products Changes in land use, settlement regulation
		Contamination of drinking water	Contamination of drinking water	Adsorptive layers Remediation Additional treatment of drinking water	Source control Use other sources than groundwater for drinking water

(*) Public welfare includes economic viability; see Table 3
(**) in this context, recreation refers to inhabitants' recreational activities

6. Stakeholder Involvement

It was mentioned at the start of this chapter that the involvement of stakeholders is necessary in order to comply with the legal requirements and in order to reduce public resistance and increase tolerance towards risks. As Caddy and Vergeze put it: "Engaging *stakeholders* in policy making is a sound investment and a core element of good governance [88]. It allows governments to tap wider sources of information, perspectives and potential solutions, and improves the quality of decisions reached. Equally important, it contributes to building trust in government, raising the quality of democracy and strengthening civic capacity." For a further elaboration on the value of stakeholder involvement in sediment management, please refer to Chapter 7 - *Risk Perception and Communication* in this book and to the book "Sediment management at the river basin scale", edited by Phil Owens, this book series. The notions of stakeholders and public are often used confusedly. 'Stakeholders' are defined as people or organised groups of people who have an effect on or are affected by sediment management (based on [89]). This definition includes amongst others lay people, environmental pressure groups, but also companies, semi-governmental organisations, and (decentralised) governments. In other words: a group of stakeholders can be diverse with regard to professional and educational backgrounds, interests and rate of organisation. 'Public' addresses a large and broad group such as the citizens of a country. Although the public can be included in a group of stakeholders once they have an effect on or are affected by sediment management, the stakeholders are not by definition the same as the public. Since sediment management and the subsequent risks do not cover everyone in a society, it is better to speak of stakeholders.

Once a policy maker decides to involve stakeholders in the process, the next step is to identify the important requirements for stakeholder involvement, and the initial design of this process. The first step towards an interactive process is to determine the goal of the process, and consequently the degree of influence assigned to the stakeholders. This is shown in Table 7.

Note that the level of participation chosen holds consequences for both the policy makers and the experts. It should be stressed that the decisions concerning the degree of participation must be communicated towards the stakeholders so as to avoid disappointments on their part. Disappointments may rise when stakeholders expect to have a higher degree of influence on the process than intended by the process manager. Formally, the stakeholders will not be allowed to have as much influence as they would like to have, which in turn might result in disappointment, cynicism and decline of support. Additionally, once a certain degree of stakeholder influence is chosen, this should not be changed to a lower degree in the course of the process, because this would provoke the same response as when the initial degree of influence is not communicated and agreed upon.

Next comes stakeholder selection. As mentioned previously, the stakeholders can comprise different groups of people, but not all people. The definition used in this section (see above) gives indications as to whom to select. Some groups must be selected and approached actively, whereas others will involve themselves. The differences between the site-specific and river basin scale level are important to keep in mind. If dealing with a site-specific situation, the stakeholders identified will be different than when dealing with sediment management on the river basin scale. In the first case, this will mean involving stakeholders such as individual farmers, local companies and interests groups and citizens that are directly affected by the sediment management. At a river basin scale, on the other hand, stakeholders such as national farmer organisations, branch organisations and national environmental organisations are more likely partners in the process. However, in both cases it is very important that although calculated risk may be low, a person living next to a sediment depot will be particularly conscious that a risk *exists*: it does not matter if the risks are calculated to be high or low. This is because of the existence of subjective risk perception as discussed in Chapter 7 - *Risk Perception and Communication*.

After determining the initial design of the process and selecting the appropriate stakeholders, the next step is to choose a medium through which the process can be run. This includes organising workshops, community gatherings, means of communication and so on (see table 7). It is important to involve the stakeholders from the start of the process. If they are involved in a later stage, this might suggest that they are not taken seriously, and they will have very little influence on the choices that have been made. Disappointed stakeholders will surely use their obstructive power and the process will become frustrated. The case of the Ingensche Waarden in the Netherlands (see Box 3) shows the negative impact of stakeholder obstruction on the progress of the process.

Box 3: Ingensche Waarden: resistance against deposition

The Ingensche Waarden is a pristine area in the Province of Gelderland (the Netherlands), located along the Rhine. In 1999, a company called the Ingensche Waarden targeted this area for deposition of contaminated dredged material. The Ingensche Waarden include a gravel pit and the idea of the company was to enlarge the pit, mine the remaining gravel and use the enlarged pit for storage of dredged material. The company applied for the necessary permits with the authorities. Operations of this scale require an environmental impact assessment, or 'milieu-effect rapportage' (m.e.r.) in Dutch.

Once it became known that the company was planning to store contaminated dredged material, resistance from local stakeholders rose quickly and the company met fierce opposition. Stakeholders organised themselves in a pressure group called 'Save the Betuwe' (the Betuwe is the name of the larger area). They organised meetings, issued publications and started to persuade local politicians to turn around the decision. One of the local citizens was a famous television personality and her involvement in the pressure group rose attention from the media. The stakeholders did not like the idea of storing contaminated sediments in an area that looked natural and did not believe the claims by the company that the contamination would remain inside the depot and would not seep into the groundwater. In turn, the company used the project reports to claim that there were uncertainties regarding the effects of the contamination on the surroundings but that these uncertainties did not involve irresponsible risks to the environment and public health.

However, the stakeholders did not believe the company and simply stated that a calculated risk does not mean anything to them. They simply did not want any contaminated dredged material in their area. As of 2005, a final decision has still not been made. This example shows how exclusion of stakeholders can have a negative impact on the progress of the decision-making process and how strong perceptions of risks are, even if calculated risks show that storing contaminated dredged material is feasible (see also Chapter 7 - *Risk Perception and Communication*).

Table 7: Degrees of influence and approaches (From: [1])

	Degree of influence	Governance style within the scale of participation.	Methods of participation Site-specific	River basin	Methods of communication
1	Stakeholders are not involved	Closed authoritarian	None	None	None
2	Stakeholders are informed – they remain passive	Open authoritarian	Use media available on a local scale	Use media available on a regional or national scale	Brochures, newsletters, internet sites, speeches, commercials
3	Stakeholders are consulted	Consulting style	Organising open evenings	Invite organizations to a discussion round, based on a stakeholder analysis	Group sessions, study groups, public gatherings, interviews, internet discussion.
4	Stakeholders give advice	Participative style	Invite stakeholders to a give advice, discuss questions/issues that needs answers. Use of local knowledge.	Invite representative organisations to an advice round, based on a stakeholder analysis	Debating meetings, house calls, internet discussions, public gatherings with discussion.
5A	Stakeholders become co-producers (Two governance styles (5A and 5B)	Delegating style	Invite stakeholders to express their concerns and give them the opportunity to suggest ideas/solutions	Invite representative organisations to an advice round, based on a stakeholder analysis	Debating meetings, house calls, internet discussions, public gatherings with discussion.
5B	can be distinguished here, with different roles of stakeholders, experts and policy makers)	Co-operative style			
6	Stakeholders do not only produce solutions but also decide about them	Facilitating style	Organizing workshops, create a common ground for discussion and decision-making, e.g. joint fact finding. Mediation if necessary.	Organizing workshops, create a common ground for discussion and decision-making, e.g. joint fact finding. Mediation if necessary.	Debating meetings, house calls, internet discussions, public gatherings with discussion.

Table 7:   Degrees of influence and approaches (From: [1]) (cont.)

	Role of the stakeholder	Role of the expert	Role of the policy maker	Potential influence on the process
1	None	Delivers information to the policy makers on demand; no information to stakeholders	Policy makers determine policy; policy process is closed, no information is issued	Initially quick, delay with implementation
2	Stakeholders receive information but don't deliver input to the process	Delivers information to the stakeholders on demand of the policy makers	Policy makers determine policy; information is issued to the stakeholders	
3	Stakeholders are consulted, act as interlocutors	Delivers information to the participants at request of all parties; experts provide another flow of information to the process, next to the flow of the stakeholders	Policy makers determine the policy and open the process to input by stakeholders, but are not obliged to adopt their recommendations	
4	Stakeholders become advisors to the process	Delivers information to all parties at request of all parties and investigate suggestions from participants on demand of the policy makers	Policy process is open to input (other ideas, suggestions etc.) by stakeholders; their input is taken into account, but policy makers have the right to deviate from it in their decisions	
5A	Co-decision makers within the set of preconditions	Experts treat policy makers and stakeholders as equal clients; advice and knowledge provision to both actors	Policy makers make take the input of stakeholders into account, and honor it, if it fits into the set of preconditions	
5B	Policy partners on the basis of equivalence	Experts treat stakeholders as equal knowledge providers; must keep an open mind to suggestions and ideas from the stakeholders	Policy makers interact with stakeholders on the basis of equivalence, they take the input of stakeholders very seriously	
6	Taking initiatives, making decisions	Experts support stakeholders with knowledge; experts treat stakeholders as their clients, need no approval of the policy makers	Offers support (money, time of civil servants, etc.) and leaves the production of solutions and decisions to the participants	Initially slow, faster implementation because of stakeholders' support

Involving stakeholders is not easy and besides values, it also involves risks with regard to the decision-making process as put forward by Gerrits and Edelenbos [1].

These pitfalls include amongst others:

Asymmetry: This characteristic covers several risks that are inherent to the disparity of stakeholders. Asymmetry in stakeholder involvement rises when some parties have an advantage over other parties. In that case, there is a latent danger that the actor who does not own a certain advantage may be overruled. At the same time, all parties usually have some kind of advantage but not in the same area. Therefore, it would be a rash conclusion to state that all actors should be equal. That would not benefit the process. Asymmetries in stakeholder involvement include lack of stakeholder representativeness, a knowledge gap as not all stakeholders have the same level of knowledge, different interests and the lack of communication [1].

Clashing expectations: Too often participants expect different outcomes than the decision actually taken. Gerrits and Edelenbos explain: 'For example, a governing body of a river can invite people living near a dredged material disposal site to come up with new ideas about how to address the disposal of contaminated sediments. They are consulted, asked to give a recommendation. However, should this not be properly communicated, the invitees might expect that they are expected to take part in the decision-making. The result will be that their expectations rise too high, and cannot be met, resulting in distrust, downright pessimism and obstruction of the process' [1].

Stakeholder out of sight: Unfortunately, there often appears to be a discrepancy between involving stakeholders in a process and the actual formal decision-making. The process of stakeholder involvement is then regarded as a way to pacify the opposition, when the actual decision mainly serves the interests of the formal decision maker, much to the disappointment of the stakeholders. Decision makers who engage in a process of stakeholder involvement should commit themselves to the process, whatever the outcomes, and not regard it is a means to lessen the opposition.

In the case of sediment on the river basin scale, cross-boundary co-operation represents a range of pitfalls for stakeholder involvement. Sediment management will often be cross-boundary and attention must therefore be paid to the characteristics of international co-operation in stakeholder processes. There are two main dimensions to such processes, which are cultural differences and institutional differences. Cultural differences are dealt with in Chapter 7: *Risk Perception and Communication*. Institutional differences (amongst others) comprise the distribution of power and authority, assigned budgets and different stages of adopting and implementing national and international legislation.

Ignoring the afore-mentioned pitfalls can have serious consequences on the process. People who are not taken seriously will be disappointed, opt-out and use their obstructive power. In conclusion, the involvement of stakeholders can be a fruitful process that can give new insights and save time and money. But this process should be approached carefully, and should be taken seriously which includes creating enough resources and time to conduct the process properly.

An example that clarifies the application of the proposed approach to a complicated site-specific management problem with potential consequences on the river basin scale is presented in Box 4: the Iffezheim case. This case has become a political issue between Germany and the Netherlands.

Box 4: The Iffezheim case

Iffezheim is the last of 12 barrages along the Upper Rhine. Its sediment is highly contaminated with hexachlorobenzene (HCB), which was discharged as an industrial by-product into the Rhine by the a company in Rheinfelden in Switzerland until the plant was closed in 1985. Since then, HCB, which strongly adsorbs to the coarse particles of the sediment, is transported downstream and accumulates in the various barrages, becoming resuspended during flood events. Until 2003, contaminated material that needed to be dredged at the Iffezheim barrage was disposed of near the barrage in containment facilities. These are now full and the authority responsible, the Waterway and Shipping Administration (WSA) had to find a way to accommodate the 130,000 m^3 of sediment that reach Iffezheim and accumulate there each year. The mass increase over time would threaten dam stability. The WSA looked for the least expensive solution at the site and after an extensive decision making process came to the conclusion that this would be the relocation of the upper most sediment layers, which show lower HCB concentrations, into the Rhine.

As HCB-contaminated suspended matter travels downstream the Rhine after flood-induced resuspension, the Port of Rotterdam in the Netherlands, which is located at the mouth of the Rhine, feared that any deliberate relocation of HCB-contaminated material into the Rhine would lead to an exceedance of the HCB threshold values in sediments in the port. This would forbid the port management to relocate the material at sea, which is the least expensive option for the Port of Rotterdam. Accordingly the Dutch stakeholders intervened, asking their German counterparts not to start the relocation process. Hereby, the Port of Rotterdam was not primarily following environmental objectives. For them, despite increases in HCB concentrations, an increase in threshold values that would allow them to relocate the material at sea would also be an option. This, however, is prevented by national and international regulations. Organisations from Germany and the Netherlands oppose the relocation plans because they will increase the Rhine's HCB-content downstream of Iffezheim. Their objective is to ensure environmental quality. Although HCB is potentially carcinogenic, strongly persistent and accumulates in the food chain, the impact on human health is probably low. A deliberate relocation of a substance, declared as one of 12 persistent organic pollutants by the UNEP, which emissions are supposed to be reduced, however, is against environmental ethics and against any European attempts to manage the environment in a sustainable way.

According to the current discussion, the German decision to relocate the material complies with national regulations. It contradicts, however, several EU principles and directives (non-deterioration principle, sustainability principle, SEA directive). The objective to meet regulatory criteria on an international level is here overridden by national compliance and the need to find a short-term solution in order to secure dam safety and to decide on an economical option.

In this example, site-specific, short-term objectives conflict with other site-specific, short-term objectives (Port of Rotterdam), which are here in line with long-term river basin objectives. All objectives are by themselves understandable and legitimate. This does not have to be the case for the preferred measures. A sustainable solution can only be achieved by accepting and understanding the different objectives of the stakeholders, by communicating the risks involved, and by trying to find the best possible option to comply with the objectives as much as possible. Henceforth it is necessary to involve different stakeholders, as has been described in section 6.

In cases like Iffezheim, when a former polluter can no longer be held liable, but previous activities provide a problem for the entire river basin, the responsibilities as well as the financial investments need to be shared. Otherwise sustainable management is not possible, as it is seldom the least expensive option. To involve all stakeholders would also mean to involve all countries that border the Rhine downstream of the original source of pollution. Switzerland, where the HCB was discharged, France managing some of the barrages upstream and where HCB-contaminated material also accumulates and is resuspended and transported downstream, as well as Germany and the Netherlands, will need to get together if this truly transboundary problem is to be solved along the lines of basin-wide sediment management. First steps have been taken: An HCB monitoring programme has been agreed upon and a new joint working group on sediments has been founded within the ICPR.

7. Conclusions

As mentioned in the introduction, the objective of this chapter was to discuss the relationships between social and societal driving forces, pressures and responses in sediment risk management and introduce indicators as triggers for choosing management options on a site-specific basis, as well as considering the effects on the river basin. By this approach, the complex system of objectives that govern and steer sediment management decisions can be structured and communication can be facilitated between different parties that may be concerned by decisions and their outcomes. Objectives can be followed with a strong site-specific or river basin-specific attitude, which is why we included indicators and management options on both scales. Furthermore, the time-scales of management objectives are different: while economic viability has a short-term and long-term perspective, ensuring environmental quality requires efforts over a long period of time. This complicates finding a common solution in a group of stakeholders with different objectives.

In this chapter, all objectives are treated in the same way, communicating the belief that no single objective is wrong and all should be taken into account. Only by understanding what drives the single person, what his or her interests are and what risk to the personal objective this person has to face, can compromises be found and agreements reached.

For the sake of clarity, social and societal driving forces, risks, indicators and management options assigned to the different objectives were completely separated. This may be an oversimplification, resulting in overlaps in several chapters. It must also be remembered that the assumption that a person follows only one objective is mostly in error. Sticking to the example above, the harbour manager who lives in the area is not only concerned with maintaining economic viability, but also wants to preserve sites with recreational quality which he may enjoy during the weekends. The local stakeholders value their recreational sites and have a strong concern for securing quality of life, which is also indirectly influenced by the economic viability of the area and hence employment and social well-being.

The objectives of different stakeholders overlap less, as the issues are broadened from a site-specific to a river basin-specific perspective, because the impacts are less direct and the perceived involvement in processes upstream that do not directly influence one's own quality of life is lower.

References

1. Gerrits L, Edelenbos J (2004): Management of sediments through stakeholder involvement - the risks and value of engaging stakeholders when looking for solutions for sediment-related problems. JSS - J Soils & Sediments 213: 239-246
2. Salomons W, Brils J (Eds.). 2004. Contaminated Sediments in European River Basins. European Sediment Research Network SedNet. EC Contract No. EVKI-CT-2001-20002, Key Action 1.4.1 Abatement of Water Pollution from Contaminated Land, Landfills and Sediments. TNO Den Helder/The Netherlands
3. WBGU (1999): Welt im Wandel: Strategien zur Bewältigung globaler Umweltrisiken - Zusammenfassung für Entscheidungsträger. Wissenschaftlicher Beirat der Bundesregierung Globale Umweltveränderungen. 25 pp.
4. Bell S, Morse S (1999): Sustainability Indicators - measuring the immeasurable. Earthscan: London. 175
5. OECD (2004): OECD Key Environmental Indicators. OECD. 38 pp.
6. UNEP/RIVM (1994): An overview of Enviromental Indicators: State of the art and perspectives. UNEP.
7. WRI (1995). Environmental Indicators: A systematic approach to measuring and reporting on environmental policy performance in the context of sustainable development. World Resources Institute: Washington D.C.
8. Smeets E, Weterings R (1999): Environmental Indicators: Typology and Overview. European Environment Agency. 19 pp.
9. Slobodkin LB (1994): The connection between single species and ecosystems. In: Sutcliffe DW (Ed.), Water quality and stress indicators in marine and freshwater ecosystems: Linking levels of organization (individuals, populations, communities): 75-87. Freshwater Biological Association, Ambleside, UK
10. UNEP (2002): Global Environmental Outlook 3. Earthscan. 446
11. Eckloff D(1992): Vibrio Species, Pseudomonas Aeruginosa und Staphylococcus Aureus in Strandbereichen von Nord- und Ostsee. Dissertation, Universität Kiel, 99 pp.
12. Venkateswaran K, Kiiyukia C, Nakanishi K, Nakano H, Matsuda O, Hashimoto H (1990): The role of sinking particles in the overwintering process of *Vibrio parahaemolyticus* in a marine environment. FEMS Microbiol. Ecol. 73: 159-166
13. Gauthier MJ, Munro PM, Flatau GN, Clement RL, Breittmayer VA (1993): New prospects on adaptation of enteric bacteria in marine environments. Mar. Life 3(1-2): 1-18
14. Heise S, Reichardt W (1991): Anaerobic starvation survival of marine bacteria. Kieler Meeresforsch. Sonderh. 8: 97-101
15. Lewis DL, Gattie DK (1991): The Ecology of Quiescent Microbes. ASM News 57(1): 27-32
16. Morita RY (1993): Starvation-Survival Strategies in Bacteria. In: Kemp PF, Sherr BF, Sherr EB, Cole JJ (Eds.), Handbook of Methods in Aquatic Microbial Ecology: 441-445. Lewis Publishers, Boca Raton
17. Magarinos B, Romalde JL, Barja JL, Toranzo AE (1994): Evidence of a dormant but infective state of the fish pathogen Pasteurella piscicida in seawater and sediment. Appl. Environ. Microbiol. 60(1): 180-186
18. Pearce R, Turner RK (1990): Economics of Natural Resources and the Environment: Harvester Wheatseaf.
19. Schuijt K (2001): The Economic Value of Lost Natural Functions of the Rhine River Basin - Costs of Human Development of the Rhine River Basin Ecosystem. Erasmus Center for Sustainable Development and Management (ESM), Erasmus University Rotterdam.

20. Renes G, Bouma GM, Wijnen W, Puylaert HJM (2003): Ruimtelijke kwaliteit in de MKBA Ruimte voor de Rivier (in Dutch). TNO.
21. Millenium Ecosystem Assessment (2005): Ecosystems and Human Well-Being: Wetlands and Water. Synthesis. World Resources Institute: Washington, DC. 69 pp.
22. Stanley Foundation (1971). Sixth Conference on the United Nations of the Next Decade. Conference held 20-29 June 1971, Sianai, Romania:
23. Carson R (1962): Silent Spring. Houghton Mifflin.
24. Meadows D, Meadows D (1972): The Limits to Growth: A report for the Club of Romes's Project on the Predicament of Mankind..
25. US Government (1980): Entering the Twenty-first Century: The Global 2000 Report.
26. Barkham J (1995): Ecosystem management and environmental ethics. In: O'Riordan T (Ed.), Environmental Science for Environmental Management: 30-44. Longman Singapore Publishers, Singapore
27. RIZA (2004): De publieke beleving en waardering van schone waterbodems en biodiversiteit in Nederland. RIZA report 2004.002.
28. Hinrichsen D (1987): Our common future. A reader's guide. The "Brundtland report" explained. Earthscan Publications Ltd.: Washington DC. 38 pp.
29. Berz G (2001): Insuring against catastrophe. Our Planet 12(1): 19-20
30. Goodess CM, Palutikof JP (1990): Western European Climate Scenarios in a High Greenhouse Gas World and Agricultural Impacts. Developments in Hydrobiology 57: 23-32
31. Schirmer M, Schuchardt B (1993): Klimaänderungen und ihre Folgen für den Küstenraum: Impaktfeld Ästuar. In: Schellnhuber H-J, Sterr H (Eds.), Klimaänderung und Küste: 244-259. Springer-Verlag, Berlin
32. Hygum B, Madgwick J, Vanderbeeken M, Blincoe P, (Editors) (2001): WWF Water and Wetland Index Assessment of 16 European Countries – Phase 1 Results – April 2001. WWF. April 2001. 72 pp.
33. FAO (2004): The state of world fisheries and aquaculture. Publishing Management Service. 154 pp.
34. Schallenberg M, Kalff J, Rasmussen JB (1989): Solutions to Problems in Enumerating Sediment Bacteria by Direct Counts. Appl. Environ. Microbiol. 55(5): 1214-1219
35. Luna GM, Manini E, Danovaro R (2002): Large Fraction of Dead and Inactive Bacteria in Coastal Marine Sediments: Comparison of Protocols for Determination and Ecological Significance. Appl. Environ. Microbiol. 68(7): 3509-3513
36. Meyer-Reil L-A (1983): Benthic response to sedimentation events during autumn to spring at a shallow water station in the Western Kiel Bight. II. Analysis of benthic bacterial populations. Mar. Biol. 77: 247-256
37. Hobbie JE, Daley RJ, Jasper S (1977): Use of nuclepore filters for counting bacteria by fluorescence microscopy. Appl. Environ. Microbiol. 33(5): 1225-1228
38. Fenchel TM (1978): The ecology of micro- and meiobenthos. Ann. Rev. Ecol. Syst. 9: 99-121
39. Fenchel T (1992): What can ecologists learn from microbes - life beneath a square centimeter of sediment surface. Functional Ecology 6: 499-507
40. Mare MF (1942): A study of a marine benthic community with special references to the micro-organisms. J. Mar. Bio. Ass. U.K. 25: 517-554
41. Wood PJ, Armitage PD (1997): Biological effects of fine sediment in the lotic environment. Environmental Management 21(2): 203-217
42. Bruton MN (1985): The effects of suspensoids on fish. Hydrobiologia 125: 221-241
43. Chapman PM, Fairbrother A, Brown D (1998): A critical evaluation of safety (uncertainty) factors for ecological risk assessment. Environmental Toxicology and Chemistry 17(1): 99-108

44. Alabaster JS, Lloyd RL (1980): Water quality criteria for freshwater fish. Butterworths: London. 297 pp.
45. Doeg TJ, Koehn JD (1994): Effects of draining and desilting a small weir on downstream fish and macroinvertebrates. Regulated Rivers: Research and Management 9: 263-278
46. Richards C, Bacon KL (1994): Influence of fine sediment on macroinvertebrate colonization of surface and hyporheic stream substrates. Great Basin Naturalist 54: 106-113
47. Rosenberg DM, Wiens AP (1978): Effects of sediment addition on macrobenthic invertebrates in a northern Canadian stream. Water Research 12: 753-763
48. Lemly AD (1982): Modification of benthic insect communitites in polluted streams: Combined effects of sedimentation and nutrient enrichment. Hydrobiologia 87: 229-245
49. Aldridge DW, Payne BS, Miller AC (1987): The effects if intermittent exposure to suspended solids and turbulence on three species of freshwater mussel. Environmental Pollution 45: 17-28
50. Lewis K (1973): The effect of suspended coal particles on the life forms of the aquatic moss *Eurhynchium riparioides* (Hedw.). I. The gametophyte plant. Freshwater Biology 3: 251-257
51. UBA (2005): Stark erhöhte Hexachlorcyclohexan (HCH)-Werte in Fischen aus Mulde und Elbe. Presseveröffentlichung August 2005
52. Heise S, Claus E, Heininger P, Krämer T, Krüger F, Schwartz R, Förstner U (2005): Studie zur Schadstoffbelastung der Sedimente im Elbeeinzugsgebiet. Commissioned by the Hamburg Port Authority.: Hamburg. 181 pp.
53. Sommerwerk K (2003): Die Recherche der industriehistorischen Entwicklung der Region Bitterfeld-Wolfen als Voraussetzung für ein effektives Umweltmonitoring. In: Zabel H-U (Ed.), Theoretische Grundlagen und Ansätze einer nachhaltigen Umweltwirtschaft: 111-130, Halle
54. Alexander M (2000): Aging, Bioavailability, and Overestimation of Risk form Environmental Pollutants. Environmental Science & Technology 34(20): 4259-4391
55. Reid BJ, Jones KC, Semple KT (2000): Bioavailability of persistent organic pollutants in soils and sediments - a perspective on mechanisms, consequences and assessment. Environmental Pollution 108: 103-112
56. Wick LY, Springael D, Harms H (2001): Bacterial Strategies to Improve the Bioavailability of Hydrophobic Organic Pollutants. In: Stegmann R, Brunner G, Calmano W, Matz G (Eds.), Treatment of contaminated soil: 203-218. Springer, Berlin
57. Gandrass J, Eberhardt R (2001): "New" substances - substances to watch. In: Gandrass J, Salomons W (Eds.), Dredged Material in the Port of Rotterdam - Interface between Rhine Catchment Area and North Sea: 289-305. GKSS Research Centre, Geesthacht, Germany
58. Nixon S, Trent Z, Marcuello C, Lallana C (2003): Europe's water: An indicator-based assessment. European Environmental Agency. 99 pp.
59. Friedrich G, Schiller W, Pohlmann M, Schwenke B, Seuter S (2000): Rhein und Nebenflüsse. In: Ministerium für Umwelt und Naturschutz LuVdLN-W, Nordrhein-Westfalen L (Eds.), Gewässergütebericht 2000 - 30 Jahre Biologische Gewässerüberwachung in Nordrhein-Westfalen: 55-78
60. Kaushik NK, Hynes HBN (1971): The fate of dead leaves that fall into streams. Archiv fur Hydrobiologie 68: 465-515
61. Wallace JB, Grubaugh JW, Whiles MR (1996): Biotic indices and stream ecosystem processes:Results from an experimental study. Ecological Applications 6: 140-151
62. Heard SB, Richardson JS (1995): Shredder-collector facilitation in stream detrital food webs: Is there enough evidence? Oikos 72: 359-366
63. Pomeroy LR (1974): The ocean's food web, a changing paradigm. BioScience 24: 499-504

64. Azam F, Smith DC, Stewart GF, Hagström Å (1993): Bacteria-organic matter coupling and its significance for oceanic carbon cycling. Microbial Ecology 28: 167-179

65. Manini E, Fiordelmondo C, Gambi C, Pusceddu A, Danovaro R (2003): Benthic microbial loop functioning in coastal lagoons: a comparative approach. Oceanol Acta 26: 27–38

66. Richardson DJ (2000): Bacterial respiration: a flexible process for a changing environment. Microbiology 146: 551–571

67. Lavelle P, Dugdale R, Scholes R, Berhe AA, Carpenter E, Codispoti L, Izac AM, Lemoalle J, Luizao F, Scholes M, Treguer P, Ward B, Etchevers J, Tiessen H (2005): Chapter 12 – Nutrient Cycling. In: Hassan R, Scholes R, Ash N (Eds.), Millennium Ecosystem Assessment. Vol. 1, Ecosystems and Human Well-Being: 331-353. Island Press, Washington

68. Galloway JN, Dentener FJ, Capone DG, Boyer EW, Howarth RW, Seitzinger SP, Asner GP, Cleveland CC, Green PA, Holland EA, Karl DM, Michaels AF, Porter JH, Townsend AR, Vörösmarty CJ (2004): Nitrogen cycles: past, present, and future. Biogeochemistry 70: 153-226

69. Tannenbaum LV (2005): A critical assessment of the ecological risk assessment process: A review of misapplied concepts. Integrated Environmental Assessment and Management 1(1): 66-72

70. Chung N, Alexander M (1998): Differences in sequestration and bioavailability of organic compounds aged in dissimilar soils. Environ. Sci. Techn. 32: 855-860

71. Heise S, Ahlf W (2002): The need for new concepts in risk management of sediments. J Soils & Sediments 2(1): 4-8

72. Burton (Jr.) GA (2001): Moving beyond sediment quality values and simple laboratory toxicity tests. Setac Globe 2: 26-27

73. Miranda ML, Mohai P, Bus J, charnley G, Dorward-King EJ, Foster P, Munns (Jr) WR (2002): Policy concepts and applications. In: Guilio RTD, Benson WH (Eds.), Interconnections between Human Health and Ecological Integrity: 15-41. Society of Environmental Toxicology and Chemistry, Pensacola (Florida), USA

74. European Environmental Agency (2003): Mapping the impacts of recent natural disasters and technological accidents in Europe. 48 pp.

75. Estrela T, Menéndez M, Dimas M, Marcuello C, Rees G, Cole G, Weber K, Grath J, Leonard J, Ovesen NB, Fehér J, Consult V (2001): Sustainable water use in Europe. Part 3: Extreme hydrological events: floods and droughts. European Environmental Agency. 84 pp.

76. European Environmental Agency (1995): Europe's environment: The Dobris assessment. European Environment Agency.

77. Tulla JG (1825): Über die Rectification de Rheins, von seinem Austritt aus der Schweiz bis zu seinem Eintritt in das Großherzogtum Hessen. Karlsruhe.

78. Kunz E (1982): Flußbauliche Maßnahmen am Oberrhein von Tulla bis heute mit ihren Auswirkungen. In: Hailer N (Ed.), Nateru und Landshcaft am Oberrhein: Versuch einer bilanz, Referate und Aussprachen der Arbeitstagung 1977.: 34-50. Veröff. Pfälz. Ges. Förderung Wiss., Speyer

79. Dröge B, Engel H, Gölz E (1993): Entwicklung und Beobachtung der Sohlenerosion im Längsverlauf des Rheins. Bundesanstalt für Gewässerkunde. 15.11.96. 22 pp.

80. LUA (2002): Hochwasserabflüsse bestimmter Jährlichkeit HQ_T an den Pegeln des Rheins. Landesumweltamt Nordrhein-Westfalen. 96 pp.

81. LUA (2000): Umwelt NRW - Daten und Fakten. Landesumweltamt Nordrhein-Westfalen. 431 pp.

82. Hajat S, Ebi KL, Kovats S, Menne B, Edwards S, Haines A (2003): The human health consequences of flooding in Europe and the implications for public health: a review of the evidence. Applied Environmental Science and Public Health 1(1): 13-21

83. USEPA (2003): EPAs Draft Report on the Environment 2003: Chapter 4 Human Health. 64 pp.
84. Ministerium für Umwelt und Forsten (2005): Merkblatt für Angler im Mosel-Saar-Einzugsgebiet. Published April 2005.
85. EUROSTAT (2005): Water abstractions in Europe. European Environmental Agency. Copenhagen, 123 pp.
86. Scheidleder A, Grath J, Winkler G, U. Stärk, Koreimann C, Gmeiner C, Nixon S, Casillas J, Gravesen P, Leonard J, Elvira M (1999): Groundwater quality and quantity in Europe. European Environment Agency. 123 pp.
87. Stec S, Casey-Lefkowitz S (2000): The Aarhus Convention - an implementation guide. UN: New York and Geneva. 198 pp.
88. Caddy J, Vergez C (2001): Citizens as Partners: Information, Consultation and Public Participation in Policy Making, report prepared for Organisation for Economic Cooperation and Development (OECD).
89. Susskind L, McKearnan S, Thomas-Larner J (Eds.). 1999. The Consensus Building Handbook. Thousand Oaks, CA: Sage.

Strategic Framework for Managing Sediment Risk at the Basin and Site-Specific Scale

Sabine Apitz[a,b], Claudio Carlon[c], Amy Oen[d] and Sue White[b]

[a]SEA Environmental Decisions, 1 South Cottages, The Ford, Little Hadham, Hertfordshire SG11 2AT, United Kingdom.
[b]Institute of Water and Environment, Cranfield University, Silsoe, Bedfordshire, MK45 4DT, United Kingdom.
[c]Consorzio Venezia Ricerche, via della Libertà 5/12, Marghera, 30175 Venezia, Italy.
[d]Department of Environmental Engineering, Norwegian Geotechnical Institute, P.O. Box 3930 Ullevaal Stadion, N-0806 Oslo, Norway.

1. Introduction

1.1. The Interdependency of Sediment Quality, Quantity and Risk

Whilst details of definitions for sediment differ, the critical factor that defines a particle as sediment (as opposed to soil, fill, etc.) is the past, present or future relationship between a particle and water. Sediments are particles that interact with, move within and are deposited below and within water bodies. Because waters are continuously dynamic, their associated particles are moving as well, resulting in a cycle of erosion, suspension and deposition throughout the natural (and anthropogenically altered) environment. It can be argued that sediment itself is of no interest as the object of management unless its presence or absence impedes an objective of society such as maintaining waterways, reservoirs, bridges, fisheries, and so on. However, its dynamic nature has required that, to maintain various water-body and land-based functions, man has for centuries undertaken to move sediments from places where water deposits them to places it does not. Such sediment management can be termed sediment quantity management. Historically, partly through a lack of environmental understanding, and partly because dredging practices were established long before the industrial revolution resulted in widespread contamination of sediment, questions about sediment quality were not generally considered. More recently, however, growing environmental and human health concerns, and restrictions on dredged material disposal have made sediment quality a dominating factor in sediment management. Currently, much of the thinking on sediment management and sediment risk assessment is dominated by concerns over sediment quality. However, even when quality is seen as a dominant concern, sediment quantity issues can also be critical in creating risk and in determining management options. The interdependence between the management of sediment quantity and sediment quality has not been effectively addressed in most risk assessment and management frameworks.

Sediment quality can be defined as the ability of sediment to support a healthy benthic population (the organisms that live in intimate contact with sediments at the bottom of water bodies). Quality can be affected by a number of physical, chemical and biological factors, but the focus of the discussion in this chapter is on sediment quality as a function of the presence (and associated bioavailability) or absence (or non-availability) of toxic chemicals. Quality can be assessed by a number of methods, including combinations of chemical measurements (often compared to standards or benchmarks), toxicity tests and/or benthic community analyses. A summary of many of these measures can be found Barcelo & Petrovic [1]. Because many contaminants have a tendency to associate with sediment particles [2,3], they can accumulate in, and be transported by sediments throughout a river basin.

Sediments are an essential part of the aquatic ecosystem, providing habitat and substrate for a variety of organisms, as well as playing a vital role in the hydrological cycle. Sediment quantity can, however, represent a risk to the well-being of a system, through imbalances, or through incompatible physical characteristics. Examples of risk include excesses or lack of sediments in rivers, estuaries, reservoirs, lakes and impoundments which can reduce storage and flow capacity, increase flood potential, damage hydro-power installations, degrade habitats, erode river channels downstream of sediment "blockages", and undermine the stability of channels and infrastructure (e.g. erosion of bridge piers). Examples of benefits include sediment supply to the nearshore environment (with implications for longshore drift/coastal stability), the provision or sustenance of wetland and aquatic habitats, sediment extraction for use in building/road industries, and beneficial use/capping of contaminants.

By thinking about sediments holistically, and at the river-basin scale, we may also need to consider some non-traditional sediment "contaminant", risk-related issues. For instance, many river banks and flood banks are contaminated with historical industrial waste or even dredged material. During flood events, contaminated sediment deposited on fields may make fields unsuitable for agricultural use. Sediments with high nutrient loads may be vital for sustaining habitats or agricultural production [4], or, in contrast, they may prove a threat to fragile floodplain vegetation communities [5]. Nutrients bound to sediments may play an important role in eutrophication [6], and pesticides and pharmaceuticals bound to sediments may prove to be a chronic problem [7,8]. It is important to remember that contaminants that are currently associated with given sediments might have originated from a different source than those sediments, and these sources may need to be managed differently.

Sediments are part of a hydrodynamic continuum. If we first consider a simple linear – and unidirectional - continuum of sediments through a system, then under the action of gravity, sediments will tend to move down the "energy gradient", i.e. from land to river to estuary and to the ocean. This is complicated at many points by flooding (river to land), tides (ocean to estuary/river), removal and deposition (estuary, river, and ocean to land or ocean), and discontinuities (dams, barrages, weirs). Therefore, sediment transfers are not always linear, or unidirectional.

Various natural processes connect sediments (and contaminants, soil, water and biota) within the hydrodynamic system. However, actions we carry out within a river basin, whether on sediments themselves or on other materials and processes within their catchment area, also affect other sites within the hydrodynamic continuum. Thus, if we take action at one point in this continuum, then we would expect the main impacts to be downstream – or at least down the energy gradient, although there may also be indirect implications elsewhere. We can use this simple conceptualisation to think about the risks, benefits, or interactions associated with different aspects of sediment in three ways:

- The removal or addition of sediment *quantity* at one point in the continuum, whether by natural or anthropogenic actions, can have an impact on sediment *quantity* elsewhere in the continuum. This can be termed a *quantity-quantity* interaction. Examples include the removal of sediment upstream potentially causing erosion downstream, or climate- or land use-related changes in river flow patterns and/or sediment supply upstream changing erosional and depositional patterns downstream.
- Upstream changes in sediment *quantity*, whether due to management actions or natural processes, can have an impact upon sediment *quality* elsewhere. This can be termed a *quantity-quality* interaction. Examples include the removal of sediment upstream causing erosion or exposure of contaminated sediment downstream, or the burial of contaminated sediments via the deposition of clean sediments.
- Changes in sediment *quality* upstream, whether by remediation, removal, source control or natural processes, can impact *quality* elsewhere. This interaction can be termed a *quality-quality* interaction. One example of this would be the removal of contaminated sediment or contaminant sources upstream reducing the risk of contaminated sediment deposition downstream.

Whilst these three types of interactions are not mutually exclusive, such a conceptual approach may provide a vocabulary or scheme for classifying different types of risks (or benefits) associated with different types of sediment-contaminant interactions in a river basin as a whole. That is, holistic sediment management must consider mass transfers from source to sink, both for sediment quantity and for the masses of contaminants associated with the sediment. Thus, a river basin-wide sediment management plan must consider in its assessment and management scheme both the solid matter *and* contaminants that can plausibly enter the sediment cycle.

There is thus a need to broaden thinking on sediment management and risk assessment to a basin-wide approach that addresses both *quantity* and *quality* of sediment, as well as a need to recognise the fact that actions or changes in conditions in one part of the basin, whether the result of anthropogenic or natural processes, may well have impacts elsewhere in the basin. Sediment quantity and quality issues must be considered together, not separately as entirely different issues, as the interdependence of quality and quantity in river basins demands a holistic approach, precluding a decoupling of these facets.

1.2. The Inter-Relationships Between Soil, Sediment, Water, Contaminants and Biota

Many organic and inorganic contaminants have a tendency to partition from the dissolved phase to the fine-grained, high surface area, organic-rich particles that make up a large portion of both bed and suspended sediments. Other contaminants already within or bound to the particulate phase enter the river system continuously. As such, whilst surface waters generally exhibit a transitory chemical signature indicative of current conditions, sediments can, in contrast, retain and integrate the chemical signature of decades to centuries (as can upland soils, waste deposits, agricultural and industrial sites and mining areas, all of which can ultimately then influence sediments). Although sediments can remove contaminants from the water, attenuating the short-term effects of human activities on the ecosystem, they can also retain a contaminant signature, possibly affecting the water column, or the organisms in contact with the sediments, for an extended period of time. Sediments reflecting past and present point and diffuse pollution in rivers and depositional areas such as reservoirs, harbours, lakes, barrages and flood plains are susceptible to erosion and further transport downstream. Furthermore, continuing agricultural and industrial practices, as well as catastrophic spills, accidents, and changes in erosional and depositional patterns due to climate change and anthropogenic activities continue to provide both point and diffuse sources of sediment, both contaminated and uncontaminated, into many river basins.

Contaminant inputs to and effects on sediment quality come from various sources and types, and via various pathways. As noted above, "properly functioning" river systems, in both ecosystem and socio-economic terms, are dependent upon a proper balance of the aspects of sediment quality and sediment quantity. Both an excess and a lack of sediments, either due to past, present or future natural or anthropogenic processes, can put various functions of a river at risk. Thus, in a river basin, both sediment and contaminant sources are manifold and their respective locations, potential source strength (a function of sediment and contaminant quantity stored) and amenability to erosion, under current and projected conditions, must be determined. A description and inventory, whether conceptual or quantitative, of the mass flow of water, contaminants and particles (and thus risk) within a river basin can be termed a Conceptual Basin Model, or CBM (see section 4.1).

Of course, as discussed above, contaminants can partition, transfer and move through a dynamic river ecosystem through various media, including air, sediment, soil, water and biota. Management of risk in such an ecosystem, or within a given river basin, suggests that sediment management should be integrated into water and soil management. Therefore, an integrated approach to soil, sediment, water and biota should be developed. Although the need for an integrated approach is well recognised and invoked at both the regulatory and scientific level, a focus on a single ecosystem component at a time is, in actual practice, usually carried out in order to avoid an overwhelming complexity. In this regard, the European Water Framework Directive (WFD), which is focused on water quality issues and recommends use of a basin-scale approach, does not address the sediment issue at depth. Similarly, the European Habitats Directive, which focuses on habitat conservation issues, does not properly address sediment and water quality at a river basin scale. Thus, a conceptual appraisal of any proposed sediment management framework in light of water- and biota-focused perspectives is required. A decision-making hierarchy, which encompasses priority setting at a basin scale down through site-specific risk assessment at a local level, is a necessary approach for managing water, soil, sediment and biota, as well as point and diffuse contaminant sources. This is entirely in line with the philosophy and requirements of the WFD.

1.3. The Need for Basin-Scale Management

Effective and sustainable management strategies must focus on the entire sediment cycle, rather than on one site at a time (often called a unit of sediment in discussions). The mission of SedNet (the demand-driven European Sediment Research Network) is "to be a European network for environmentally, socially and economically viable practices of sediment management at the river basin scale" [9]. One of the main aims of SedNet is to "develop a document containing recommendations in the form of guidance and key-solutions for integrated, sustainable sediment management (SSM), from local to river basin scale (SSM Guide)" [10]. SedNet is not the first organisation to develop sediment management guidance, and a number of reviews of sediment management approaches are available [11,12], but some of the goals and drivers for SedNet differ from those for other published guidance. Firstly, as an EC-sponsored program, SedNet addressed a number of international and cross-border issues not always addressed elsewhere. Secondly, in line with the new regulatory focus of the Water Framework Directive (WFD), SedNet addresses river basin-wide management, and basin-wide sediment management in particular. Thirdly, although most guidance documents have been generated for specific aspects of sediment management (such as dredged material disposal or environmental management [11] a basin-scale approach must integrate various sediment goals and provide an universal framework. Different nations, organisations and stakeholders have different objectives when they address sediments, and frameworks must be devised that allow goals and priorities to be balanced in a transparent way. The goal of SSM demands that sediments are managed, not one unit of sediment at a time, but instead with the interactions between that unit and all current or potential sources or sinks within a river basin, in mind [13].

One of the main drivers for European river basin management and for the SedNet initiative is the European Commission's WFD. The WFD (Annex VII) requires member nations to develop River Basin Management Plans (RBMPs) which are required to deliver the WFD objective of good, or improving, ecological status in all water bodies. These RBMPs will need to consider many aspects of basin-scale management within the socio-economic environment of the region, country and continent. The concept of basin-scale *sediment* management, which will be discussed in this chapter, is just one facet of such an RBMP, with other facets including water resources, flooding, nutrient management, priority substances and biota. As with sediment management, many of these other facets will also require understanding of the hydrodynamic continuum, and thus any plan derived for sediment management must be compatible with other requirements of the RBMP. Sediment in and of itself is most closely related to the good hydromorphological status required by WFD, although the potential for sediment to bind, transport and/or store contaminants means that sediment may also play an important role in determining whether or not we can meet the targets of good chemical and biological status over the short to medium term. "Good" in this context should be thought of in reference to ecological status, and thus chemical, biological and hydromorphological status are required in support of ecological status rather than as ends in themselves. Moreover, the baselines against which chemical, biological and hydromorphological status will be judged will differ from site to site and are still in the process of definition. Ecological status expresses the quality of the structure and functioning of aquatic ecosystems associated with surface waters, following the classification expressed in Appendix V of the WFD. This appendix does not provide a general definition for each type of organism (phytoplankton, macrophytes and phytobenthos, benthic fauna and invertebrates etc.). For all groups however there is a common frame: "excellent" status is obtained if the taxonomic composition and abundance do not differ substantially from reference conditions and "good" corresponds to slight modifications in either composition or abundance.

Because sediment management has generally been fragmented by different people who manage sediments for different reasons, many people use the same terms or phrases for different steps in the sediment management process. As one example of this, sediment assessment and management frameworks designed to support decisions related to dredged material disposal are fundamentally different from those designed to determine whether contaminated sediments pose a human health or ecological risk *in situ*. For one thing, these frameworks for decision making are applied at very different points in a decision process. In the former, a management decision has already been made (to dredge) and the "assessment" focuses on the selection of an appropriate methodology and location for reuse or disposal. In the latter, questions are being asked about whether risk exists, and thus the "assessment" focuses on whether there is a need to manage sediments at all. Different methods, assays and assumptions apply, but it is often difficult to clarify these differences, since similar vocabularies are applied.

Therefore, to achieve sustainable sediment management in a river basin, the various practitioners of sediment management must come to the table before any sediment management decisions are made and develop the sediment-specific aspects of RBMPs that will balance the environmental, economic, social and regulatory needs throughout the basin. To achieve such a balance, a common language must be developed such that priorities can be established and understood by all parties involved, information needs can be defined and filled, and sediment can be managed in a sustainable way.

The complex, multi-scale and multivariate nature of holistic sediment management requires the involvement of many "layers" of political, technical, scientific, economic and environmental analysis, which can be difficult to integrate and unify. Whilst many of these processes involve very different drivers, organisations and approaches (whether on site-specific or river-basin scale), holistic sediment management requires that the relationships between these processes, including their points of interaction, intersection and information exchange, be clearly defined (see Chapter 2). The dynamic nature of river sediments and the often international aspects of sediment-related problems call for a new approach to sediment management in which transport, quantity and quality are explicitly addressed throughout the assessment, decision making and management processes. To achieve these goals, it is necessary to develop strategic, conceptual and process frameworks that identify these interactions, define common issues and terms, and help define and thus expedite information exchange in support of effective sediment management.

2. Frameworks for Basin-Scale Management

2.1. Basin-Scale versus site-specific assessment and management

The UK Ministry of Agriculture, Fisheries and Food (MAFF) has produced a series of documents discussing strategic planning for flood and coastal protection (defence) in which specific projects or schemes are evaluated in terms of large-scale planning and strategy development [14]. These documents state:

"For the purpose of large-scale planning, project appraisal can be used, over a wide area and taking a broad approach, sufficient to build a guiding framework within which layers of smaller-scale strategies or schemes can be developed. Similarly for individual scheme development, the appraisal process can ensure that the most suitable option is selected and progressed. At each level, all the potential impacts and options are considered to an appropriate level of detail and geographical scale to ensure good decision making and option selection. ... the ideal planning structure is frequently referred to as a strategic framework, indicating that each stage of planning has a context in a wider defined picture."

It follows that any flood defence scheme (or any project that might impact upon flood defence) must be conceived of in terms of strategic goals for the entire shoreline management unit. This approach is somewhat analogous to the management of specific aspects of a river basin within terms of a larger basin management plan, inasmuch as any site-specific management action within a river basin should be assessed in terms of its role and effect on basin-wide plans and objectives.

Apitz and White proposed a conceptual framework to address how sediments are managed at basin-wide and site-specific scales [15]. However, evolving discussions within SedNet and other groups since then have suggested a number of refinements to the framework. Firstly, there is a need to recognise the fact that not all sediment decisions should, or can, start at the basin-scale (in advance of site-specific consideration), even if such is the ideal. Thus, information must flow in both directions - site-specific studies can help guide basin-scale evaluations just as basin-scale evaluations should inform site prioritisation. Secondly, source control must be more explicitly address within the framework. Thirdly, it must be made clearer that risk assessment is only one part of the decision process and that sediments are only one medium that must be managed in a basin. Lastly, there was a need to clarify that site prioritisation (which is based upon risk in this framework) should be only one aspect of the development of the Basin Use Plan, now called the River Basin Management Plan (RBMP) in this document.

Figure 1 portrays a simplified conceptual framework defining the relationship between basin-scale and site-specific considerations in river basin management. Note that this diagram need not be sediment-specific or risk-specific, but rather can be used to address management options for various media. A more detailed, sediment risk-specific process diagram that fits into this framework will be presented in Section 3. As described above, management of river basins will require an evaluation of all relevant processes (both natural and anthropogenic) at the basin-scale using all available data. Whilst the focus of this chapter and book is on sediment risk management, such can only be achieved within the context of its broader context. However, it is important to reiterate that the selection, prioritisation, implementation and evaluation of any management action is dependent upon a broad range of factors (including, but not limited to, economic appraisal, technical assessment, risk assessment and environmental assessment), all of which must be evaluated at every level of the decision-making and management process. Figure 2, adapted from the MAFF approach [14], shows the various types of appraisals necessary at each step in the decision process, and how information from each feeds into other steps. Clearly, every step of a decision step must consider budgets, cost-effectiveness and cost benefits (economic appraisals); technical feasibility (technical assessment); and all aspects of risk, including regulatory and socio-economic risks (risk appraisal), as well as the ecological risks and benefits of any action. Thus, whilst this paper will focus on ecological risk (or environmental) appraisals, it should not be forgotten that these are only one part of the appraisal and decision-making process.

Evaluation at the basin scale may result in the elimination of some proposed management actions. However, for those specific actions, projects or control measures identified as a result of a RBMP, site-specific assessment must be carried out, followed as appropriate by management and monitoring.

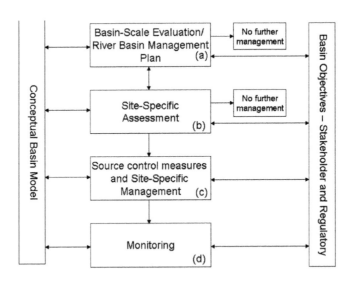

Figure 1: Conceptual diagram of the relationship between basin-scale and site-specific assessment and management in a river basin.

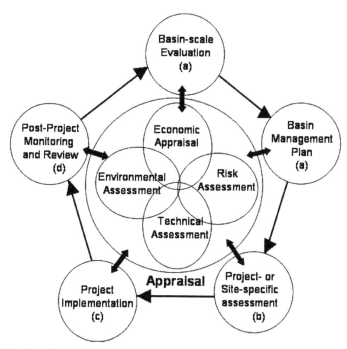

Figure 2: The iterative appraisal process at various phases of a basin scale/site-specific evaluation process (adapted from [14]). Note a, b, c and d in circles correspond to process steps in Figure 1.

2.2. Selection of Site-Specific or Basin-Scale Initial Approach for a Given Project

There are a number of reasons to manage sediments (as well as water, rivers and ecosystems) holistically, and at the basin scale, rather than site-by-site or unit–by-unit. Firstly, to do so is explicitly required by the WFD. If each site in a river basin is managed in isolation, managers are often reacting to symptoms rather than looking at causes and thus, in a dynamic system, site-by-site management can only be a short-term solution. It makes sense to manage in a manner that prioritises sites that pose risk to other sites, and to do so in a manner that considers the entire sediment and contaminant budget, from source to sink. Impacts and benefits of management actions for purposes other than risk reduction (e.g., waterway maintenance, wetland development, etc.) should be evaluated in concert with the remedial and socio-economic goals for a given basin. Such an approach will provide insight into the changes in agricultural, industrial and development practices that will most effectively reduce sediment and contaminant inputs, maximise the potential for beneficial use and hence will reduce the cost of maintaining waterways and protecting the environment. This approach will also help ensure that the various goals and objectives for a river basin can be met with minimum impact on other objectives within the basin. This is a route to sustainable development. Such an approach will help focus limited resources to maximise the achievement of basin management objectives. While these objectives include basin-scale risk reduction, the more traditional drivers for sediment management, such as navigational dredging, are also more effectively implemented when basin-scale interactions are addressed. In this approach, we explicitly address the compromises that are implicit in decisions we make about river basin management.

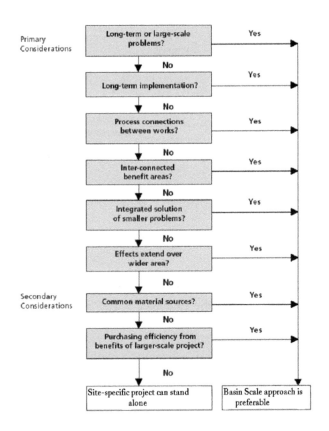

Figure 3: Decision tree for site-specific vs. basin-scale management actions (adapted from MAFF, 2001).

There will, of course, be cases in which all aspects of a specific sediment management action are local, for which it will not make sense to additionally consider the complexity of basin-scale issues. Figure 3 illustrates a number of questions that should be asked to determine whether it makes sense to carry out a site-specific management activity as a stand-alone project, or whether river basin-scale strategic issues should first be considered, at least at some level. Because of the dynamic nature of river systems, the use of Figure 3 questions will often make clear the utility of basin-scale considerations, even for site-specific problems. However, in those cases in which sediment management decisions begin at the site-specific level (e.g., dredging for construction, small cleanup projects, etc.), it is essential that the impacts of these actions upon adjacent sites, and on basin-scale quantity, quality and objectives are also considered, and that information and decisions made at the site-specific level continue to be communicated to those who act at the basin-scale.

3. Sediment Risk Management at the Basin and Site-Specific Scale

Despite recent calls for river management at the basin-scale, most sediment assessment and management frameworks developed thus far only address sediment at a site-specific, rather than a basin, scale [11,12]. One way to address this inconsistency is to devise a framework that considers sediment and contaminant sources to a river basin relative to one another, balances these issues with socio-economic goals for a basin, allows a place for sediments in a basin management plan, and then feeds into site-specific risk analysis. Using such an approach, sediments are first evaluated at a basin scale, then site-specifically. Figure 1 presents such a framework, not just for sediments, but also for management of all relevant media considered at the river-basin scale. Because evaluation at the basin scale may initially occur at a screening level, or may be based upon generic criteria, risk assessment is needed for those sites designated as potentially high risk in a basin-level assessment, in order to inform and refine conservative risk assumptions using site-specific conditions. Site-specific assessment may suggest that some sites proposed at the basin level for management do not require further action, or may suggest that the cost or risk of a proposed project such as a dredging action is too risky or expensive to carry out. All remaining proposed actions, whether conducted for the purpose of risk reduction or to fulfil socio-economic objectives, should undergo an evaluation of the risks, consequences and benefits of the particular proposed action. After a thorough evaluation of all proposed actions, remaining site-specific management actions, including source-control measures, should be conducted, and then followed by appropriate monitoring. Results of monitoring can be used to refine Conceptual Basin Models (CBM, see Section 4.1), as well as provide for appropriate modifications to the RBMP.

Clearly, some of the details regarding the manner in which this approach can be implemented for specific media and appraisal issues must still be clarified. Specifically, this section will discuss how various aspects of risk appraisal fit into a basin-scale/site-specific framework, with an emphasis on sediment-related issues.

With regard to sediment management, risk should be assessed at two different spatial scales: the basin scale and the local (site-specific) scale, as well as at various stages in the decision-making process. Though some of the fundamental factors being evaluated may be the same for both scales (such as risk to human health, the environment and to river basin objectives), the methods, degree of detail and information available will differ. For example, prioritising sites or management actions at the basin scale involves setting priorities for sediment units throughout the basin that account for several kinds of risk and management objectives, and prioritises these sediment units for both socio-economic and ecological management actions. Whilst Apitz and White (2003) initially defined this process as "Site Prioritisation", subsequent evaluations and discussions have made clear that this term encompassed too many disparate variables to be addressed in only a risk-focused appraisal. Thus, the term "Risk Prioritisation" will be used here and also in Chapter 4, to describe the basin-scale risk-relevant factors that can be used to prioritise sites in a basin in terms of relative risk, thus replacing the above initial term. This, then, separates the risk factors from other factors, while still recognising that many other technical, socio-economic and regulatory factors, all outside the scope of this paper, will be brought to bear before sites are ultimately prioritised for the sediment-specific aspects of an RBMP. In this Chapter, the term sediment Basin Management Plan (sBMP) will be used to refer to these sediment-specific aspects that will inform the RBMP or other basin-scale management activities. This will be further discussed in Section 4.3.

Because of the extensive analyses required to provide sufficient information to prioritise site risks at a basin-scale, much of the data used in the analysis will necessarily be based upon screening-level information, as well as literature or generic criteria. Thus, sites that are given a high-priority status based upon screening-level risk evaluation will require further, site-specific and detailed analysis of risk before being subjected to potentially costly management actions. These sites are thus subject to site-specific risk assessment which can be defined as the evaluation of individual sediment parcels to determine and rank their risk relative to benchmarks, site- or basin-specific criteria. A further risk evaluation at the local scale (although with a significant basin-scale component) is needed to characterise risk to the environment for any given management option (dredging, for example), or to compare several options. This risk evaluation can be termed Project Risk Appraisal. When several options are being evaluated, this is often termed a Comparative Risk Assessment.

Figure 4 presents a process diagram for basin-scale and site-specific risk management, which details the risk appraisal aspects of the conceptual framework in Figure 1. The letters denoting various hierarchical levels of the diagram (Figure 4) correspond to the same processes in Figure 1. This diagram is comprised of the three principal levels of risk evaluation described above.

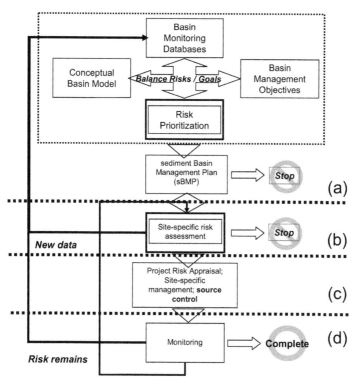

Figure 4: Process diagram for basin-scale and site-specific sediment risk management. A manager may "enter" the process at the basin scale (in level "a"), or at the site-specific scale (in level "b"). Note: whilst the focus of this figure is on sediment risk management, the same concepts could be applied to other media.

Risk appraisal is only one aspect in the development of a sBMP but it is the risk-related aspects of such a process that are emphasised in this process diagram, and which represent a risk-specific "cross-section" through Figures 1 and 2. In risk-related terms, basin-scale assessment begins with the processing and synthesis of information derived from basin monitoring and from databases to develop a Conceptual Basin Model (CBM), and an assessment of the various objectives that are desired for the river basin as a whole and for sites within the river basin in particular (Basin Management Objectives, see section 4.2 and Chapter 2 – *Sediment Management Objectives and Risk Indicators*). The CBM, which is discussed in detail in Section 4.1, considers the mass flows of particles and contaminants, and a screening level assessment of sediment quality and archived data. This information is then used to evaluate, if possible, the relative risk associated with various sites, with a consideration of the potential risk implications of those management actions being considered to achieve socio-economic goals (such as dredging, construction, etc).

A number of other appraisals (see Figure 2) should then be carried out to generate a sBMP, which should present a prioritised list of proposed projects and strategies, some to achieve ecological goals, and some, such as dredging, to support socio-economic goals. It should be pointed out that an important part of any sustainable sBMP is a contaminant and sediment source control plan. The success of such a control plan is largely dependent upon a CBM that adequately accounts for diffuse and point sources throughout the basin. Low-priority sites or projects (which can be eliminated based on appraisals) are then set aside for no further action, while relatively higher-priority sites or projects are subject to more detailed assessment. Sites given high priority based upon screening risk will undergo a detailed risk assessment, referred to herein as site-specific risk assessment. The results of such risk assessment may determine that, based upon more detailed information, management of the site based on risk is unnecessary, or the assessment may instead demonstrate the need for risk management. Site-specific risk assessment is characterised by tiered assessment and the determination of site-specific risk (see Chapters 4 and 5 for detailed discussions of site-specific risk assessment). Site-specific management options should be evaluated using a project risk appraisal, and comparative risk assessments. This approach will be used to evaluate and compare the potential risks and benefits of various remedial, management and/or disposal options based upon site-specific impacts on BMOs, site-specific risk, technical feasibility and regulations, as well as potential impacts upon other sites within the system (as informed by the CBM). Proposed projects in support of socio-economic (rather than remedial) goals will still be subject to site-specific risk assessment and a project risk appraisal, though often at a much lower level, if there is no evidence of site-specific risk. They may then, however, go straight to management for socio-economic goals without a remedial step. All projects should be subject to post-project monitoring, the results of which can both be used to evaluate the efficacy of any project as well as to help revise the CBM.

As previously noted, not all projects have broad (e.g. basin-wide) implications, and current as well as past regulatory frameworks and decision makers do not and have not acted at the basin scale. Thus, in many cases, the process diagram shown as Figure 4 will be entered into at point (b), with a site-specific risk assessment and a Project Risk Appraisal. A properly conducted Project Risk Appraisal will consider potential impacts of a given management action upon adjacent and downstream sites, and will ideally consider available upstream solutions, thereby effectively addressing some basin-scale considerations. In all cases, many aspects of these evaluations are iterative, with new information informing various levels of the process. To date, agencies, decision makers and governments have not yet developed all the policy or infrastructure for making basin-scale evaluations and decisions (although this should be changing with the advent of the WFD). Furthermore, the methodologies and procedures necessary to carry out such broad-reaching evaluations and decisions – regardless of who is involved - are still in very early stages of development. In short, implementation of a basin-scale approach will be necessary if river basins are to be managed holistically and in line with emerging policy, therefore, iterations of such an approach must first be laid out in clear fashion, then refined and debated.

4. Components of Basin-Scale and Site-Specific Sediment Management

4.1. Conceptual Basin Model and Risk Prioritisation

The flow of particles within a river system, whether clean or contaminated, can impact the relative risk, quality or potential utility elsewhere in the hydrologic continuum, e.g. at downstream sites. Furthermore, actions taken in the basin can affect sites downstream, increasing or decreasing the quality of downstream sediments, as well as altering the dynamic balance of particles. It can be argued that it is the relationship between hydrodynamically connected sites - in terms of quality, quantity and energy - that defines a site's relative risk, and its (risk-based) priority in a risk management strategy. However, site-specific decision frameworks have not generally addressed the dynamic relationship between sediment sites in a formal way. Just as a conceptual *site* model allows risk assessors to consider the flow of contaminants to target organisms in support of site-specific risk assessment [16-18], an understanding of the particle and contaminant mass flows within a river basin in support of basin-wide management and prioritisation, should inform basin-scale evaluation.

This description and inventory, whether conceptual or quantitative, of the mass flow of contaminants and particles (and thus risk) within a river basin can be termed a Conceptual Basin Model (CBM, [15]). Such a model is a critical part of effective risk assessment of a particular site, and of river basin management as a whole. The complexity of a CBM will differ from river basin to river basin, depending on the information available. It can be quite conceptual (as is the one described for illustrative purposes below), or it can involve detailed chemical, sedimentological, hydrodynamic and modelling studies. As with a conceptual *site* model, the Conceptual *Basin* Model should be frequently updated and refined as more data become available from basin-scale and site-specific assessments, and from monitoring and research. Many tools available for the development of a CBM are described in the SedNet WP2 book.

While a CBM may be very complex in reality, we use the example below to illustrate, in a simplified way, how information from a CBM can be used to inform assessments of relative risk in support of a Risk Prioritisation. To simplify matters, one can begin with the assumption that, in a basin, an energetic continuum can be defined between particles, water and contaminants - from source to ultimate sink - and that materials move along this continuum. The analogy of a "jerky conveyor belt" has been used by others [19] although traffic flow models may be more appropriate. For simplicity, one can assume that this flow is unidirectional (i.e., materials move only in one direction), although it is clear that there are discontinuities and sinks within river systems (e.g., a dam will stop sediment flow, at least over long time periods, and tidal forces can move materials upriver). If a river basin is first divided into units (or sites, or parcels, used interchangeably here) of a given size, the position of a parcel of sediment or source material can then be defined on an energetic continuum, as can a parcel's relationship to an adjacent parcel as either a sink or a source. One can also assume that sites can be (and have been) evaluated in terms of quality, at least at a screening level. The definition of quality, as used herein, will depend upon the benchmarks and objectives selected for a given basin, as is discussed below. However, for a given measure of quality, one can characterise parcels of sediment or source material relative to one another.

Then, one can map the sediments in a basin in terms of their energetic position (source to sink) and (screening level) quality. Figure 5 illustrates a conceptual diagram of a projection of sediment energy (source vs. sink) and quality. The x and y axes represent latitude and longitude. The z axis on the bottom graph represents the energetic state of a parcel of sediment (in a simple case, this represents elevation at grid points in the basin, as particles move from the source to the oceans). Whilst consideration of energetic state in terms of potential energy sounds very elegant, other factors such as hydrodynamics, slope, erosive energy, shear stress, landscape complexity, and so on, will likely also help predict particle mobility.

In this example, *potential* sediment is also included because, in a dynamic system, assessment should be broadened to include not only those materials that are currently sediments, but also materials such as soil, mine tailings, etc. that can reasonably be expected to become part of the sediment cycle during the timeframe of basin-scale management.

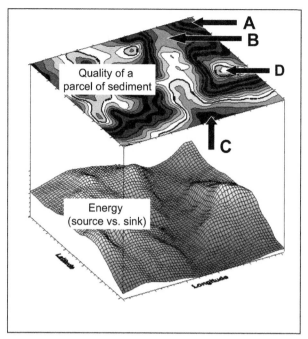

Figure 5: Conceptual diagram of a projection of sediment energy (source vs. sink) and quality using data from a CBM to inform Risk Prioritisation.

The frequency with which various areas of a river basin will contribute sediment to the aquatic system will differ greatly. Some areas will deliver particles and/or contaminants on a regular basis whilst others will only deliver during extreme events. Delivery patterns and frequency may alter as a function of climate change as well as land management changes.

The example provided in Figure 5 is an oversimplification of reality in many ways. One oversimplification is that the entire basin surface is included in the assessment whereas, in a real risk map, there will be many areas of the basin that do not represent either current sediments or plausible source materials for future sediments. Nevertheless, this map is useful for illustrative purposes, and can be refined when case study data become available.

Figure 5 can also be used to consider a series of sediment parcels within the same river system. The grey-scale projection at the top of the figure is a contour plot of sediment quality (based on a hypothetical chemical or biological measure). In this figure, poorer quality sediment is indicated by black and dark grey, better quality grades to white. As can be seen, sediment unit A is black (poor quality) and has a high elevation or energy, in absolute terms. More importantly, though, is its relationship to the sediment site adjacent to it, site B. Because A has both higher energy and poorer quality than unit B, it can be considered high-priority sediment, as it not only represents potential in-place risk, but also a potential risk to adjacent sites. This can be contrasted with site C. Sediment site C is also black (poor quality), but it is at a lower energetic level than the adjacent site D; thus, though potential in situ risk exists, transport risk is relatively low. It should also be pointed out that since the energetically upstream sediment site D is of relatively high quality, the prioritisation of C can be considered even lower, as the clean sediment upstream, D, if left in place, may, over time, move and attenuate the risk at C by burying it. Clearly, to be done correctly, such sediment site assessments must also consider erosive potential, as sediments may be very stable, with a low probability of moving on their own. That said, energy within a river basin changes seasonally, as a result of climate variability, and as a result of anthropogenic activities. Thus, currently quiescent sediments or potential sediments may not remain so.

Based on the above conceptual discussion, one can conclude that the relative Risk Prioritisation for this example site would be: A>C>B≥D. It should be pointed out that, with further analysis, myriad other factors could change these priorities. Nevertheless, this approach allows for sediments to be prioritised in common terms, relative to one another, for the purposes of determining order of risk management and allocation of resources. As sediment goals are refined, this model can either be simplified or adapted for greater complexity. Of course, objectives for sediment management other than risk reduction must also be considered and balanced as well, as discussed further below, and in Chapter 2.

For clarification, the above discussion assumes that there is a minimum, and uniform, amount of hydrodynamic and quality information available on sediments throughout the example basin that allows for some first-order decisions about the relative priority of these parcels of sediment. Quantity and mass flow issues must be addressed in a basin-scale assessment. There are a number of hydrodynamic models available to predict mass flow of particles. GIS-supported models and databases developed in support of water quality studies (such as SWAT [20], ANSWERS [21], MIKE-SHE [22] and the European Chemical Industries Research Council (CEFIC) Long-range Research Initiative (LRI) models such as GREAT-ER [23], TERRACE [24] and GEMCO[1] might provide a springboard for how to map mass flow of particles and contaminants in river basins. The generic requirements of a methodology to determine both potential source areas and potential transport rates include an understanding of the current location and characteristics of sediment, soil and contaminants as well as a representation of the hydrological system, including climate inputs and terrain. In practice, such information is often difficult to obtain, and whilst topographic, soil and vegetation maps (together with rainfall and temperature data) may be available, there are many gaps to fill, in both data

---

[1] http://www.cefic-lri.org/files/EventDocs/SETAC2003_Gemco_abs.pdf

availability and knowledge, that will require targeted study. Of particular note in this regard is the lack of explicit understanding of the role that landscape complexity plays in the movement of water, soil and contaminants to river systems, and the current inability to model the dynamics of supply and transfer of sediment from its full range of fluvial, rural and urban sources. These issues can be considered and addressed as problems of spatial and temporal scale. Such problems are not unique to sediment or contaminants, and it is important that modelling is able to address all issues which require an understanding of basin-level dynamics and the implications of change (e.g. water resources, flooding, priority substances).

A particular lack in terms of understanding sediment-related risk are data on soil or sediment quality at the basin level. Information used to evaluate sediment quality will depend upon the definition (and measure) of ecological quality, and management goals for a basin. Both chemical measures and toxicity assays have strengths and weaknesses as preliminary screening tools, but it is not feasible to apply all methods to all sediments in a regional assessment. Indicators of sediment quality will depend on management objectives (see Chapter 2) and may include: concentrations of specific contaminants (in absolute terms or relative to selected benchmarks or sediment quality guidelines), eco-toxicological indicators, measures of ecological health or indicators of bio-accumulation [16,17,25]. Information on these measures can also be found in Barceló & Petrovic [1].

For basin-scale prioritisation, some uniform measures of sediment quality must be selected and applied throughout the basin. The measures must be sensitive enough to detect potential risks, broad-based enough to allow for a comparison of sites with different contaminant types (unless the management goal is focused on only one or a few contaminants) and should also provide a broad dynamic range. In short, sediments with different levels of potential risk should be distinguishable from one another. Thus, output of the evaluation process will not be purely binary (i.e., pass/fail), but should also indicate relative degree of potential risk. Indicators of sediment quality for such a basin-scale assessment should be sufficient to provide preliminary prioritisation of sites, but simple and/or inexpensive enough for wide, cost-effective coverage. For data-rich systems, desk-top studies may sometimes be sufficient. For a basin-wide assessment, screening-level measures may be appropriate. Some countries, river basins and regions have already initiated such regional assessments (e.g., [26], although many others have not.

Any sediment-related risk designation made by such broad-based, screening-level assessments should only be considered as a preliminary indicator of potential or screening risk. Areas considered to be of potential concern must be confirmed with further, detailed, site-specific risk assessment. Further, it is widely recognised that proper use of multiple types of data ensures a better understanding and assessment of a site's ecological status, and is preferred over considering only a single parameter (chemical, for example) or a narrow set of parameters, even if such parameters are considered highly accurate. When considering different types of evidence, a Weight of Evidence (WoE) approach is usually invoked [27-29]. Details of such approaches are addressed in Chapter 4 and in Barceló & Petrovic [1].

4.2. Setting Basin Management Objectives

It is not only components of the CBM that will drive a prioritisation of sites and management actions. Societies also have a number of socio-economic goals for sediments and river basins which include regulatory, economic, aesthetic, recreational and ecological factors (see Chapter 2 for a detailed discussion). In order to sustain economies, evaluate, prioritise and improve ecological status of sediments, stakeholders must decide on management objectives, and how they are to be measured and balanced. The Water Framework Directive mandates "Good ecological status of water bodies" (WFD, 2000/60/EG). Those who work with sediments have no trouble understanding and communicating the link between sediments and the ecological status of water bodies, but the specific definition and measure of "good ecological status" in sediments are still being refined throughout Europe. The EC Habitats Directive mandates that there is a "Duty to demonstrate no harm" (Council Directive 92/43/EEC). Whether this "duty" applies to future, ongoing or planned anthropogenic activities, the consequences of such activities, and/or to various environmental media, must still be clarified. Most international bodies, including SedNet, speak of the goal of "sustainability". SedNet defines this term as *"The multipurpose management of sediment with full attention to adverse effects, so as to enhance the utility of river basins in the future"* [30].

As an additional complicating factor, EU programmes driven by economic goals such as the Common Agricultural Policy (CAP) currently subsidise practices that may conflict with environmental sustainability. For example, wheat and maize are now being grown in the UK on highly erosive and sloping soils. Such practices are causing greater flooding, the filling of streets and gardens with sediments [31,32], and severely silted watercourses. Consequently, regional, national, international and/or basin-wide objectives must be defined. A series of questions may be posed in this regard: Is the management goal to reduce the mass balance of contaminants in a river basin, to limit the exposure of these contaminants to the food chain, to protect benthic organisms, fisheries, shipping or farmers, and/or to achieve the "right" amount of sediment of the right type for ecological requirements? How can the degree of goal achievement be measured? How do these goals fit into current (or future planned) policy or frameworks?

There must therefore be a designation of objectives for the management of a given river basin, which can be defined as Basin Management Objectives. Objectives to be met during management generally fall under the following categories: 1) meeting regulatory criteria (e.g., WFD, Habitats Directive, North Sea Treaties, National and local legislation; see Chapter 6 – *Sediment Regulations and Monitoring Programmes in Europe*), 2) maintaining economic viability (e.g., navigation, fisheries, flood control, recreation), 3) ensuring environmental quality and nature development and 4) securing quality of human life (see Chapter 2 for a discussion of these issues). To develop Basin Management Objectives, all stakeholders must come together to identify and define site-specific and regional goals for a given basin. Such stakeholders may come from many fields (regulators, dredgers, fishermen, shippers, environmentalists and the general public, among others) and all should have input. The CBM, with as much information as is currently available should provide structure and insight into how various site-specific actions taken in one basin location might affect other basin sites and their potential uses.

One way in which the interaction between these objectives and various management actions can be communicated is with the European Environment Agency DPSIR approach (DPSIR stands for Drivers, Pressures, States, Impacts and Responses). Particularly useful for policy-makers, DPSIR builds on the existing Organisation for Economic Co-operation and Development (OECD) model and offers a basis for analysing the inter-related factors that impact the environment. The aim of such an approach is threefold: 1) to be able to provide information on all of the different elements in the DPSIR chain; 2) to demonstrate interconnectedness of the elements, and 3) to estimate the effectiveness of Responses. In chapter 2, the development of BMOs, involved risks and the indicators of risk to these BMOs, are described and their interconnectivity explained at length.

Basin Management Objectives must be defined, and will ultimately control how we prioritise sites, how we assess and manage them, as well as how extensively we include an evaluation of land-based practices in a sediment management strategy. Inevitably, goals will differ from place to place and even from time to time, and not all objectives may be achieved. The question of who decides on the necessary compromises is also an important one. As the WFD mandates the involvement of all stakeholders throughout the decision process, a clearly communicated framework for defining and balancing BMOs is essential to successful management.

4.3. Sediment-Specific Aspects of a River Basin Management Plan (sBMP)

Various appraisals come together at the basin scale, which result in a balancing of the results of the Risk Prioritisation and the BMOs (all informed by the CBM). The end-product of this balance is the development of an sBMP. The methodologies used for balancing these various factors are outside the scope of this book; although much work has been done in this regard and is on-going, these issues still require considerable work. A large part of this balancing process, however, will involve the successful interaction and communication between various agencies, practitioners, levels of government and the public. Thus, effective communication of risk and other factors amongst these entities is vital. Throughout the process of developing an sBMP, designing subsequent investigations, and planning for management activities, the involvement of all stakeholders in the sediment assessment and management process is viewed by many as critical [33]. According to the U. S. National Research Council [34], the most successful sediment management projects they evaluated involved the broadest range of project stakeholders, both early and often. Stakeholder groups can include representatives from local communities and governments, fishermen, industries, ports, environmental and public interest groups, as well as regulatory and trustee organizations from local, state and national organizations. This broad representation allows all interested parties to be involved with and understand the problems, their investigation and resolution, fostering trust and the development of a consensus, if possible [33,35]. A discussion of risk-perception and communication can be found in Chapter 7 of this book.

The manner in which potentially competing goals related to sediment management are balanced is not a purely technical decision. Rather, decision-making additionally requires consideration of socio-economic and political appraisals. Acceptance and implementation of the approach ultimately chosen and followed will require significant work, both technical and political. Different nations, organizations and stakeholders have different goals for sediment management. While SedNet is an open organization, and seeks input from all European nations and from academia, business, government and NGOs, representation is not uniform. Whether a single framework can successfully provide the tools for these disparate agendas and stakeholders to be reconciled remains to be seen. However, the definition of BMOs, and the subsequent development of a Risk Prioritisation and sBMP are critical parts of sustainable sediment management. This chapter does not address how these choices will be made, since significant work must still be done on defining river basin-wide and/or Europe-wide methods, goals and priorities, but rather seeks to suggest a framework within which to clarify the dialogue.

4.4. Source Identification, Classification and Control

Critical for success of river basin management is a plan for source control via appropriate prioritisation of sediment parcels as well as an evaluation of activities in the basin that either provide continuing point and diffuse input of sediment or contaminants, or somehow interfere with sediment-balance issues. To this end, the CBM and supporting data should be developed to such a level that sources, pathways of exposure and receptors are identified. Contaminated sediments may play the role of sources as well as receptors of risk, through equilibrium with water and biota as well as through sediment transport. Being a potential source of pollution, contaminated sediment parcels can be compared to other basin-scale sources of point and diffuse pollution. It should be noted that whether sediments are considered as sources or receptors of risk is also a function of the environmental objective of concern. If the objective is protection of the ecosystem and human health, contaminated sediments should be regarded as *sources*. On the other hand, the protection of sediment quality or quantity itself may lead to the consideration of sediment in terms of *a receptor*.

In order for source control to be effective, an understanding of the sources and mechanisms of sediment and contaminant inputs is necessary. There are a number of types of particle and contaminant sources within a dynamic river system. Sources of particles and contaminants can be divided into two broad classes: historical and active sources. Historical sources include point (e.g. landfills, contaminated soil) and diffuse sources (agricultural, atmospheric or infrastructure related contamination) as well as contaminated sediments (parcels of sediments which are acting or have acted as sinks of contamination in the past). Active sources include point (e.g. industrial and urban discharges, usually under control) and diffuse (e.g. agricultural, industrial and urban) sources. These sources can pose a risk to human health and the ecosystem through direct exposure routes and/or bioaccumulation/biomagnification processes. They can also pose a risk to human health and the ecosystem indirectly, through the ongoing contamination of sediment (e.g. downstream sediment parcels).

A different time perspective should be adopted for the management of historical and active sources. For source control, a number of factors should be considered. Firstly, contributions from various sources should be compared. Secondly, management of contaminated sediment should not be contemplated without a clear understanding of contaminant sources, and a plan for their control. Any type of intervention should be considered in terms of time. For instance, interventions to manage sediments which have continuing upstream contaminant sources (designated as secondary contaminant sources in Fig. 6) can decrease site-specific risk only over a short time frame, whereas interventions to manage active sources can reduce site-specific and overall basin-scale risk over the long term. For instance, if a given contaminated sediment continues to receive inputs from active, historical or secondary contaminated sediments upstream, it is likely that contamination in that sediment, and thus, the site-specific risk, will not decrease, or may even increase over time (line 1 in Figure 6). On the other hand, removal/control of active sources (line 2 in Figure 6) will result in a net reduction of risk to the downstream site, as that source will not continue to provide inputs, and the site can recover, but risks will still remain and can even increase due to continuing inputs from historical sources and secondary contaminated sediments. It is only if all sources are controlled, and a site is remediated or allowed to recover (if biogeochemistry and hydrology allow this) that risk at a site can be controlled over the long term (line 5 on Figure 6). If resources or technology do not allow for a control of all sources, it is important to ensure that management is applied where it is most effective. Furthermore, cleanup standards and objectives should reflect the consequences of these choices. For instance, it is unreasonable to expect that, for instance, a site-specific cleanup will provide risk reduction in the long term to levels lower than those expected if re-contamination from uncontrolled sources is still possible. It is an understanding of the relative risks at sites, and the potential risks to sites from various sources, that allows for an allocation of resources that will result in the maximum net risk reduction in a basin.

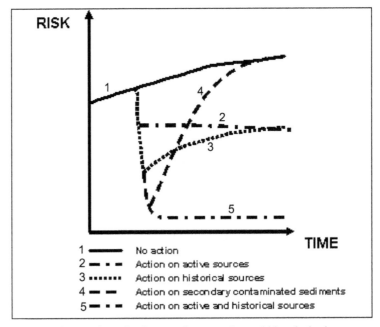

Figure 6: Risk reduction vs. time after intervention on active and historical primary sources and on secondary contaminated sediments.

4.5. Important Issues for Sediment Management Option Appraisal and Selection

Site-specific risk assessment should be carried out on sites prioritised for management. Approaches for site-specific risk assessment are addressed in detail in Chapters 4 and 5, and will not be discussed here. There are also many other site-specific risk-assessment frameworks already available in the published literature. For site-specific assessment to be meaningful within a basin-scale framework, the CBM, BMOs and the issue of source control must be considered at each step of the assessment process.

Proposed management options must be evaluated, both in terms of site-specific and basin-scale risks and benefits, in a Project Risk Appraisal; the methodologies of which are outside the scope of this chapter. However, it should be pointed out that management options for sites being managed to reduce risk as well as for low-risk sites that are prioritised to be managed to meet non-remedial BMOs must be evaluated in terms of the CBM, overall BMOs and source-control issues, and in terms of the other appraisal factors described in Figure 1. For contaminated sites, the most cost-effective solution for managing a site may be upstream source control. On the other hand, some options that address an objective at one site may have negative impacts for other sites in the basin. These issues are particularly important in river basins, because they are so dynamic. To address this, questions that should be asked and carefully considered during selection of the most appropriate options for managing contaminated sediments include:

- Is the site erosive or depositional?
- Will implementing certain management options change erosive vs. depositional status, and how will that impact other sites?
- Can sediments entering a parcel be considered as part of a solution that reduces risk, e.g. via burial, mixing or attenuation?
- Are there land-use changes planned for upstream locations that would reduce sediment loading to downstream locations?
- Does sediment entering a parcel bring with it new or more contaminants?
- Do remedial options increase risks downstream?
- Is the sediment a resource needed in the basin, e.g., in order to provide habitat or prevent channel erosion?
- Are the characteristics (e.g., grain size, organic matter content) of sediment arriving at a site appropriate to the objectives for the site?
- Is source control technologically and politically feasible?

After management actions have been selected and applied, monitoring must continue until risks are deemed to have reached acceptable levels. CBMs should be either continuously updated or periodically reviewed, and re-balanced in terms of changing BMOs and/or monitoring results. This process will support the WFD requirement of periodic reviews of RBMPs. Basin-scale sediment management should be an iterative process, and, if done properly, resources can be allocated for maximum benefit.

5. Conclusions

The proposed conceptual approach to basin-scale sediment management provides possible frameworks for addressing the complexities inherent in concurrently managing sediments at both a basin-wide and a site-specific scale. Acceptance and implementation of a basin-wide management approach will require significant work, both technical and political. Successful development of a basin-scale decision framework should provide a basis for parties with very different goals to come together in support of sustainable sediment management. Because all stakeholders are ultimately stewards of the same ecological and economic resources, breaking down technical, political, and socio-economic barriers, whether real or perceived, should ultimately help to balance the various groups' often disparate objectives for sediment management in a mutually agreeable, beneficial and sustainable way.

References

1. Barceló D, Petrovic M (Eds.). 2006. Sediment Quality and Impact Assessment of Pollutants. Elsevier: Amsterdam
2. Förstner U (1987): Sediment-associated contaminants - an overview of scientific bases for developing remedial options. Hydrobiologia 149: 221-246
3. Kersten M, Forstner U (1985): Chemical fractionation of heavy metals in anoxic estuarine and coastal sediments. submitted to Wales Science and Technology
4. Walling D, He Q, Blake W (1999): Use of Be-7 and Cs-137 measurements to document short- and medium-term rates of water-induced soil erosion on agricultural land. Water Resources Research 35(12): 3865-3874
5. Janssens F, Peeters A, Tallowin J, Bakker J, Bekker R, Fillat F, Oomes M (1998): Relationship between soil chemical factors and grassland diversity. Plant and Soil 202(1): 69-78
6. Alexander RB, Elliott AH, Shankar U, Mcbride GB (2002): Estimating the sources and transport of nutrients in the Waikato River Basin, New Zealand. Water Resour. Res. 12: 4.1-4.23
7. Brown CD, Hart A, Lewis KA, Dubus IG (2003): P-EMA (I): simulating the environmental fate of pesticides for a farm-level risk assessment system. Agronomie 23(1): 67-74
8. Vermeire T, MacPhail R, Waters M (2003): Integrated human and ecological risk assessment: A case study of organophosphorous pesticides in the environment. Hum. Ecol. Risk Assess 9(1): 343-357
9. Brils J (2002): The SedNet mission. Journal of Soils and Sediments 2(1): 2-3
10. Salomons W, Brils J (Eds.). 2004. Contaminated Sediments in European River Basins. European Sediment Research Network SedNet. EC Contract No. EVKI-CT-2001-20002, Key Action 1.4.1 Abatement of Water Pollution from Contaminated Land, Landfills and Sediments. TNO Den Helder/The Netherlands

11. Apitz SE, Power B (2002): From Risk Assessment to Sediment Management: An International Perspective. Journal of Soils and Sediments 2(2): 61-68
12. den Besten PJ, de Deckere E, Babut MP, Power B, DelValls TA, Zago C, Oen AMP, Heise S (2003): Biological effects-based sediment quality in ecological risk assessment for European waters. Journal of Soils and Sediments 3(3): 144-162
13. Förstner U (2002): Sediments and the European Water Framework Directive. Journal of Soils and Sediments 2(2): 54
14. MAFF (2001). Flood and Coastal Defence Project Appraisal Guidance. Overview (including general guidance). Ministry of Agriculture, Fisheries and Food
15. Apitz SE, White S (2003): A Conceptual Framework for River-Basin-Scale Sediment Management. Journal of Soils and Sediments 3(3): 132-138
16. Apitz SE, Davis JW, Finkelstein K, Hohreiter DL, Hoke R, Jensen RH, Kirtay VJ, Jersac J, Mack EE, Magar V, Moore D, Reible D, Stahl R (2005): Assessing and managing contaminated sediments: Part II, Evaluating risk and monitoring sediment remedy effectiveness. Integrated Environmental Assessment and Management (online-only) 1(1): e1-e14
17. Apitz SE, Davis JW, Finkelstein K, Hohreiter DL, R Hoke RHJ, V J Kirtay JJ, Mack EE, Magar V, Moore D, Reible D, Stahl R (2005): Assessing and Managing Contaminated Sediments: Part I, Developing an Effective Investigation and Risk Evaluation Strategy. Integrated Environmental Assessment and Management 1: 2-8
18. ASTM (1995): Standard Guide for Developing Conceptual Site Models for Contaminated Sites Vol. 04.08. American Society of Testing and Materials.
19. Ferguson RI (1981): Channel form and channel changes. In: Lewin J (Ed.), British Rivers: 90-125. Allen and Unwin
20. Neitsch SL, Arnold JG, Kiniry JR, Williams JR (2001): Soil and Water Assessment Tool - Theoretical Documentation Version 2000. Grassland, Soil & Water Research Laboratory, Agricultural Research Service.
21. Bouraoui F, Dillaha TA (1996): ANSWERS-2000: runoff and sediment transport model. J. Env. Eng. 122: 493-502
22. Refsgaard JC, Storm B, Refsgaard A (1995). Recent developments of the Systeme Hydrologique Europeen (SHE) towards the MIKE SHE, Symposium on Modelling and Management of Sustainable Basin-Scale Water Resource Systems (Symposium H6), XXI General Assembly of the International Union of Geodesy and Geophysics, Vol. IAHS Publ. No., 231: 427-434: Boulder, Colorado
23. Feijtel T, Boeije G, Matthies M, Young A, Morris G, Gandolfi C, Hansen B, Fox K, Matthijs E, Koch V, Schroder R, Cassani G, Schowanek D, Rosenblom J, Holt M (1998): Development of a geography-referenced regional exposure assessment tool for European rivers - GREAT-ER. J. Hazardous Materials 61: 59-65
24. Beaudin I, White SM, Hollis J, Hallett S, Worrall F, Whelan M (2002). The use of SWAT-2000 for diffuse source contaminant modelling at the regional scale across Europe, Proceedings of British Hydrological Society Symposium: 21-28: Birmingham, UK
25. Wenning RJ, Ingersoll CG (2002). Summary of the SETAC Pellston Workshop on Use of Sediment Quality Guidelines and Telated Tools for the Assessment of Contaminated Sediments, 17-22 August 2002: 44. Society of Toxicology and Chemistry (SETAC). Pensacola, FL, USA: Fairmont, Montana, USA
26. Peerboom R, van Hattum B (2000). The National Policy Framework in the Netherlands. Dredged Material in the Port of Rotterdam - Interface between Rhine Catchment Area and North Sea - Project Report. Part D, Current and Future Policies and Regulatory Framework:

27. Batley GE, Burton GA, Chapman PM, Forbes VE (2002): Uncertainties in sediment quality weight of evidence (WOE) assessments. Human and Ecological Risk Assessment 8(7): 1517-1549

28. Burton GA, Chapman PM, Smith EP (2002): Weight-of-evidence approaches for assessing ecosystem impairment. Human and Ecological Risk Assessment 8(7): 1657-1673

29. Chapman PM, McDonald BG, Lawrence GS (2002): Weight-of-evidence issues and frameworks for sediment quality (and other) assessments. Human and Ecological Risk Assessment 8(7): 1489-1515

30. Brils J, participants S (2003). The SedNet Strategy Paper, http://www.sednet.org/materiale/SedNet_SP.pdf:

31. Evans R, Boardman J (2003): Curtailment of muddy floods in the Sompting catchment, South Downs, West Sussex, southern England. Soil Use and Management 19: 223-231

32. Boardman J, Evans R, Ford J (2003): Muddy floods on the South Downs, southern England: problem and responses. Environmental Science and Policy 6(1): 69-83

33. USEPA (2002): Principles for managing contaminated sediment risks at hazardous waste sites. U.S. Environmental Protection Agency, Office of Solid Waste and Emergency Response.

34. NRC (1997): Contaminated Sediments in Ports and Waterways: Cleanup Strategies and Technologies. National Academy Press: Washington, D.C. 295

35. NRC (2001): A risk-management strategy for PCB-contaminated sediments. National Academy Press: Washington, D.C.

Prioritisation at River Basin Scale, Risk Assessment at Site-Specific Scale: Suggested Approaches

Marc Babut ([a]), Amy Oen ([b]), Henner Hollert ([c]), Sabine E. Apitz ([d,f]), Susanne Heise ([e]), and Sue White ([f])

[a]*Cemagref, Freshwater Ecosystems Biology Research Unit, Ecotoxicology - 3 bis Quai Chauveau, CP 220, 69336 LYON CEDEX 9 (France)*
[b]*Norwegian Geotechnical Institute, Sognsveien 72, 0806 Oslo (Norway)*
[c]*Department of Zoology, University of Heidelberg, Im Neuenheimer Feld 230, 69120 Heidelberg (Germany)*
[d]*SEA Environmental Decisions, 1 South Cottages, The Ford Little Hadham, Hertfordshire SG11 2AT, (United Kingdom)*
[e]*Technische Universität Hamburg Harburg, Beratungszentrum für Integriertes Sedimentmanagement (BIS), Eissendorfer Str. 40, D-21073 Hamburg (Germany)*
[f]*IWE, Cranfield University, Silsoe, Bedfordshire, MK45 4DT, (United Kingdom)*

1. Introduction

Basically, management implies making decisions; in many domains, if not most, making a decision appears difficult, firstly because different processes are involved in the problem at stake, secondly because different groups of people, having various concerns and values, are interested in the final decision. The former reason is related to complexity, whereas the latter refers to public participation.

A large array of management frameworks and methods has been developed to address the complexity issue. These approaches aim at combining heterogeneous variables (or criteria) in such a way that simple and non-ambiguous figures are related to the options available for the manager. Multicriteria Assessment methods have been thus developed for various purposes; their use in environmental management has been more recent, but is rapidly increasing. On the other hand, the public participation issue requires that additional considerations and involvement are included in the management process. Consequently, the decision process will progress by iterations between two "poles", the experts involved in the problem analysis (problem analysts) on one side and the stakeholders on the other.

In this perspective, the problem analysts should aim to provide methods to evaluate and often rank the available management options. In addition to being scientifically sound, these methods should only be as complex as necessary to address the appropriate level of complexity, and transparent enough to improve communication and facilitate public participation. Methods allowing for the categorization of data into different classes should be assessed in this perspective.

With regard to sediment management, there is a need for classification methods at two different spatial scales, the river basin scale and the local, site-specific scale. Decision-making at the river basin scale involves setting priorities for sediment units throughout the basin that account for several kinds of risk and management objectives, and prioritising these sediment units for both socio-economic and ecological management actions (see *Chapter 3 – Strategic Framework for Managing Sediment Risk at the Basin and Site-Specific Scale*). Because of the large-scale analysis required to prioritise sites at a basin-scale, much of the data used will of necessity be based upon screening-level information, and literature or generic criteria. Thus, sites that are given high priority based upon screening-level risk evaluation will need to undergo further, site-specific and detailed analysis of risk before being subjected to potentially costly management actions. These sites are thus subject to site-specific risk assessment. At this stage, if a management action is adopted, there is a range of management options available. The best management option can be selected using comparative risk assessments. Chapter 5 (*"Risk Assessment Approaches in European Countries"*) describes European approaches to sediment risk assessment in two categories: *in situ*, corresponding to site specific risk assessments, and *ex situ*, corresponding to comparative risk assessment for dredged material disposal options. Comparative risk assessment methods for remedial options are not described in this chapter.

Sediments themselves, as well as the organisations and regulatory criteria that affect them, are affected by actions at both the site-specific and river basin scale. These scales are constantly interacting through time; and the interactions are not only the consequences of the respective institutional positions and responsibilities, they also result from stakeholder (public) involvement, as summarised in figure 1. These aspects are more thoroughly discussed in Chapter 7 (*"Risk Perception and Risk Communication"*).

Risk assessment and risk management frameworks are commonly used to facilitate decision making processes [1, 2]. It can thus be considered that the risk assessment phase is a part of the management decision process. Most risk assessment frameworks include four basic elements: problem definition, analysis of effects and exposure, characterisation, and a synthesis of this information to estimate risk [3, 4][1]. In this perspective, the spatial scale at which this process occurs will influence the selection of the *assessment endpoints* [4] as well as the indicators involved at the risk characterisation step[2] and the accuracy of the risk estimates. Considering that assessing the risk at basin scale will help to determine priorities but will not lead directly to final management decisions, we will call site *prioritisation* the process aiming at defining the critical issues which should be addressed at this river basin scale. Furthermore, *comparative risk assessment* will refer to the comparison of the available management options at the site-specific scale.

This review will intentionally focus on a few selected classification, or ranking, approaches and assess their relevance in terms of public participation as well as in terms of technical or scientific aspects. Examples of how these approaches can be used for either basin scale or site-specific sediment management will be provided.

According to Burton et al. [5] there is no single "correct" semi-quantitative ranking system, as the appropriate method depends on the project objective and in this case the appropriate scale. One could probably add that the particular ranking method itself has a relatively secondary importance, provided the right expertise has been involved in the method development and its parameterisation.

---

[1] See also chapter 5
[2] See also chapter 2

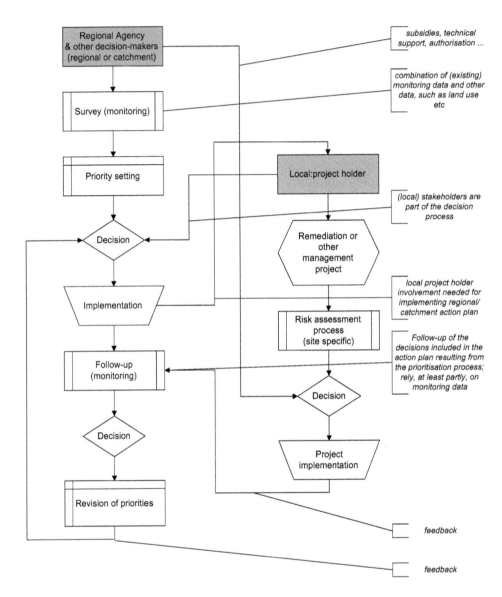

Figure 1 – Interactions of management processes at river basin and site-specific scales

2. Classification and ranking approaches

2.1. General ideas

Weight of evidence (WOE) is a term commonly used in the literature as an all-encompassing definition of a process used in environmental assessment to evaluate ecological conditions when multiple lines of evidence (LOE) are involved [5, 6]. WOE is then essentially a generic name used for any approach which combines information from more than one LOE in order to characterise risk and classify or rank management options.

G.A. Burton et al. [5] have reviewed different WOE approaches, categorising them into the following eight groups: qualitative combination, expert ranking, consensus ranking, semi-quantitative ranking, Sediment Quality Triad (SQT), broad-scale WOE, quantitative likelihood and tabular decision matrices. The authors evaluate the strengths and limitations of the different groups of WOE approaches according to five categories: *robustness*, *methodology*, *sensitivity*, *appropriateness* and *transparency*. *Robustness* is defined as the consistency in interpretation irrespective of when and where conducted. *Methodology* refers to the ease of use of the WOE approach. *Sensitivity* refers to the capacity of the method to discriminate between different levels of (adverse) effects, ranging from extreme conditions to intermediate ones. *Appropriateness* refers to whether the WOE approach is useful for a wide range of conditions or environments. *Transparency* indicates how readily understandable the approach is. Their review is summarised in Table 1.

Table 1- Advantages and limitations of different WOE approaches according to [5].

	Robustness	Methodology	Sensitivity	Appropriateness	Transparency
Qualitative Combination	Low	High	Low	High	Low
Expert Ranking	Low	Moderate	Moderate	High	Low
Consensus Ranking	Low	Moderate	Moderate	High	High
Semi-Quantitative Ranking	Low	Moderate	Moderate	High	Low
Sediment Quality Triad	Low → Moderate	Low → Moderate	High	High	Low → High
Broad-Scale WOE	Moderate	Low	Moderate	High	Moderate
Tabular Decision Matrices	Moderate	Moderate	High	High	High
Quantitative Likelihood	Moderate	Low	High	High	Moderate

As shown in Table 1, WOE approaches range from qualitative approaches like Qualitative Combination, which refer to approaches which simply combine the LOE results non-quantitatively, to quantitative approaches like Quantitative Likelihood, which include statistics to define uncertainty. Semi-quantitative or ranking approaches, intuitively lie somewhere in between the qualitative and quantitative WOE approaches. Qualitative approaches are easy to use (high level of methodology); however their conclusions are not as consistent (low level of robustness) and they do not discriminate as well between different levels of adverse effects (low level of sensitivity). On the other hand, quantitative approaches, that include likelihood and statistics to define uncertainty, can be much more difficult to use. However, such approaches are much more consistent and do discriminate between different levels of adverse effects.

According to Table 1, the only difference between expert ranking and consensus ranking procedures lies in the level of transparency. One could argue that both expert ranking and consensus ranking are approaches that occur in any WOE, whether the WOE is qualitative or quantitative. Therefore the degree, to which a WOE approach is transparent, is directly dependent on the level of stakeholder involvement. Stakeholders can participate in a number of ways, including by assisting in the development of management goals, by proposing assessment endpoints, by providing valuable insights and information, and by reviewing assessment results [3]. Classification from the viewpoint of the decision-making process is also reflected in methods such as ELECTRE [7, 8] or similar decision-aids; these methods aim at ranking various options by building a consensus among decision-makers in a case-specific context. Conversely, many approaches used for ranking chemicals, or hazard or risk of materials, focused more on the properties or characteristics of the "objects" to rank than on consensus among stakeholders. Such a strategy obviously has pros, e.g. the approach can be applied beyond the range of items (chemicals, hazardous materials, etc.) which were used for its development, and cons, among which are the frequent lack of consensus on the properties selection, or of transparency in setting boundaries.

SQT is a specific example of a semi-quantitative WOE approach and is defined by the authors as an approach that incorporates other WOE approaches (e.g. qualitative, ranking, and quantitative) to varying degrees [5]. The robustness and methodology of the SQT approach are evaluated by the authors as varying between low to moderate, whereas the sensitivity is evaluated as high. Table 1 also shows a high level of transparency for Tabular Decision Matrices. Such matrices are not an approach in themselves, but rather a means of disseminating information or presenting results. Therefore, WOE approaches that include Tabular Decision Matrices or similar ways of systematically documenting information result in a higher level of transparency

Table 1 illustrates the obvious, that when it comes to selection of a WOE approach that can be used for risk ranking, we should strive for approaches that are semi-quantitative and include a systematic method for documenting information such as Tabular Decision Matrices. In order to satisfy a high level of transparency, approaches that include consensus ranking are the most attractive. Such a stipulation will also satisfy the requirement of a high level of public participation, as stated earlier.

2.2. Classification methods

Integration within and between LOE is typically conducted using one or a combination of methods: best professional judgement (BPJ), indices, statistical summarisation, scoring systems, and logic systems [9].

Best professional judgement is an inherent element in any system, to varying degrees. Therefore, it does not need specific comment. However, it should be noted that BPJ is subject to the experience and biases of the specific practitioners. Statistical summarisation is based on descriptive methods, for instance multivariate analysis techniques; as for BPJ, they do not need specific comments.

Semi-quantitative approaches have been widely used in environmental studies, including sediment studies (several examples quoted in [5]). The few approaches proposed in this chapter were selected because they had already been applied to sediment studies, so their characteristics seemed appropriate in respect to their *robustness*, *methodology*, *sensitivity*, and *transparency*.

The use of indices or the use of a ratio to reference value has been commonly practiced when for example combining chemistry data. Grapentine et al. defined a sediment quality index based on the number of substances that exceeded sediment quality guidelines and the magnitude of exceedence [10]. In a specific case of hazard ranking of contaminated sediments, Wildhaber and Schmitt developed separate indices for chemistry, toxicity and benthic community structure [11]. Chemistry data were compared to sediment quality data and averaged. Toxicity data were compared to laboratory control values and averaged. A biotic index (mean tolerance to pollution per organism) was used to estimate relative toxicity of sediments based on the taxa abundances in the sediments. Chapman developed a graphical index (triangular plot) using data from a "triad" of assessment tools, namely chemical contaminant characterisation, laboratory-based toxicity to surrogate organisms, and indigenous biota community characterisation [12]. Data from each LOE were first normalised to the site reference conditions. This concept has since been criticised, e.g., by Chapman himself as an error of earlier studies, since there is a substantial loss of information and the significance of any spatial impact cannot be determined statistically [13]. Recent SQT studies thus tend to use sophisticated multivariate approaches for assessing individual and combined SQT components [13-17]. In that perspective, SQT should no longer be considered as an index, but rather as an assessment strategy [18].

A disadvantage of most indices is that too much mathematical manipulation of the data is carried out, and thus some information is lost. On the other hand, disadvantages of statistical methods, such as multivariate methods, are that they are more complex, and require large data sets. Furthermore, their outcomes are rather difficult to extrapolate, as they strongly depend on the dataset involved. In addition, these outcomes are not easy to interpret and understand, particularly for lay people.

Scoring systems represent a formal method for assigning values or scores to a class of measures for a particular variable. Depending on the number of variables being considered, a final score is often a mathematical algorithm based on the individual scores and ranking is accomplished by placing the objects in numerical order. For example, Wildhaber and Schmitt scaled each of their three indices from 1 to 100 and then combined all three LOE by taking the arithmetic mean of all three scaled ranks [11]. Canfield et al. also scaled variables from 1% to 100%, and ranked the results in 5 classes of 20% each [19], which is not exactly a scoring approach but provides a similar result. The "sediment quality index" developed by Grapentine et al. for the Great Lakes [10] is based on the same rationale as the Canadian Water Quality Index (CWQI); it results from a combination of two variables, namely *scope* (ratio of the number of failed variables to total number of variables), and *amplitude* (degree of non compliance). This combination is then scaled to 100. Many other systems are available, either for ranking hazards of chemicals or materials, including sediments. A common characteristic shared by these systems as well as by the two abovementioned examples is weighting; in other words, it is generally acknowledged that some variables should have more influence on the final scores than others.

Score ordination is a semi-quantitative WOE approach using scores, that has been applied in France to assess the relative risks of pesticides to surface waters, at sub-basin [20] or regional [21] spatial scales as well as for risk screening of chemicals [22, 23]. Environment Canada used this approach for assigning environmental impairment scores to local industries of the same sector [24].

Although scoring methods provide discrete results, these can be considered as quantitative and be used further in multivariate analyses, if deemed necessary. This property can for instance help in testing their reliability. However, their designs seldom account for uncertainty.

Fuzzy logic based systems depend to a larger degree on expert knowledge and BPJ. They demand the determination of variability of the different parameters and the quantification of uncertainties. In order to combine different LOE, If-Then-rules are developed which again are based upon expert knowledge. Although fuzzy systems offer a high degree of transparency because they are linguistically described (e.g. in "low", "medium" and "high" – categories and in "if … then" rules) and thus relatively easy to follow, they have been called "too subjective" and too much based on decisions by experts, thus being open to manipulation. The challenge for those applying fuzzy systems will therefore be to clearly demonstrate and communicate the basis for the implied variability and uncertainties, as well as for the rules that build up the decision system.

Application of fuzzy systems enjoys increasing popularity in technological fields [25], and their suitability for environmental assessments has also been frequently recommended e.g. for impact assessment [26, 27], for soil sciences [28, 29], for the classification of endangered species [30], for a decision analysis of polluted sites [31], or for "integrated environmental vulnerability assessment" [32], and many more, including sediments [33, 34].

3. Setting priorities at the basin scale

Sediment management at basin scale requires that various kinds of risks, some of which are not independent from each other, are addressed [35]. At this spatial scale, management objectives can be placed in four categories[3]: (i) *meeting regulatory criteria,* (ii) *maintaining economic viability,* (iii) *ensuring environmental quality and nature development and (iv) securing quality of human life.*

One way the interaction between these objectives and various management actions can be communicated is presented in detail in Chapter 2 ("*Sediment Management Objectives and Risk Indicators*"). Identification of the different management objectives, stakeholders may be following, how these are influenced by social and societal driving forces and what kind of risk they aim to reduce, will facilitate communication with policy-makers and offer a basis for analysing the inter-related factors that impact on the environment. Finding risk indicators that help measuring the occurrence and extent of a risk are supposed to help defining potential management options. In increasing the transparency of motivations and required decision-support parameters an approach such as the one developed in Chapter 2 shall: a) help providing information on all of the different elements in the decision-making process; b) to demonstrate their interconnectedness, and c) to estimate the effectiveness of management options.

In the case of sediment management, although the corresponding "drivers" differ between these four categories of objectives, risk indicators at this spatial scale tend to overlap, or to display logical connections: for instance, indicators corresponding to regulatory objectives will rely upon relatively simple criteria, such as "chemical contamination" or "indicator species", whereas the indicators related to economical risks will be broader if not more vague, for instance "decrease of fish yields" or "filling of navigation channels". Because chemical contamination in sediments impacts the management of dredged materials, or might be one of the causes of fish population losses, it can be considered that the indicator "chemical contamination" is included in the broader indicators used when applying objective (ii)[4]. Similar arguments could be developed for the other objectives. Furthermore, a sustainable sediment management strategy will necessarily address all the issues, rather than focus on a specific aspect. For that reason, and because these management objectives are indeed not

---

[3] see *Chapter 2 - Sediment Management Objectives and Risk Indicators* - for a detailed discussion of this issue

[4] discussion on indicators in chapter 3.

independent when setting management actions, a general prioritisation approach can be proposed at the basin scale, using a pre-defined set of criteria (see *Chapter 3 - Strategic Frameworks for Managing Sediment Risk at Basin and Site-Specific Scale*).

3.1. Classification criteria selection

A prioritisation approach at the basin scale will result from a compromise between an "ideal" and a realistic approach. Data availability is often the limiting factor for the realistic situation, either because historical data are used or affordability limits new data collection. At the basin scale however, the need for accurate data is somewhat limited, as only a relative or screening-level evaluation is required, as high priority sites will be subjected to site-specific, or comparative (when several management options are available), risk assessment before any management action is carried out.

An overview of the links between management targets, or drivers, and the criteria which could be used for setting priorities is given in Table 2[5].

Table 2 – Example of drivers and criteria allowing setting of priorities at basin scale

Drivers	Criteria	Comments
Navigation maintenance	Sediment budget (quantity, mobility)	Sediment excess in deposition areas Erosion, risk of bank collapse
Water quality restoration	Sediment quality	Contamination by nutrients or toxic or accumulative substances, with a potential to be released into the water column
Flood management	Sediment budget, and quality	Sediment mobility; loss of river capacity; mobile or erosive sediments may be contaminated, and may affect depositional areas
Ecological quality restoration	Sediment quality	Contamination by nutrients or toxic or accumulative substances; likelihood of contaminant transport to cleaner sites

Sediments are part of a hydrodynamic continuum; contaminant or quantity imbalances from one site can affect sites down the energy gradient, essentially on an axis from up- to downstream6. Thus, actions taken in the basin can either positively or negatively affect sites downstream as well. While site-specific risk assessment can determine the absolute ecological risk of sediments at a given

---

[5] For further details on drivers and risk indicators, see chapter 2
[6] although flows are not always unidirectional

site, the relationship between hydro-dynamically connected sediments, in terms of quality, quantity and energy, can be used to define their _relative_ risks [35]. Hydrological disconnections (e.g. dams) will modify the relative risks to some extent, but will certainly not eliminate them: hazardous sediments upstream dams could be released downstream in the case of extreme floods, or in order to restore the water storage capacity of the reservoir. As a consequence, a risk-based management prioritisation at basin scale is not only a matter of quality. The issue is to map (prioritise) the sediments (units) in a basin in terms of real and potential relative risks, and their possible interactions.

The relative risks of sites within a basin will be the result of their location within the basin, their potential energy, their sediment quality and the risk or benefit expected if a sediment moves downstream [35]. For example, a contaminated sediment located at an upstream point in the basin, and with a high potential energy, represents an in situ risk as well as a potential risk to downstream sites.

Briefly, these criteria could be described as follows; it should nevertheless be noted that these criteria are prospective (they need to be tested) and intentionally simplified. Again, the goal is not to describe and quantify thoroughly and accurately the potential risks, but to provide a screening level risk prioritisation of sediment units within the basin.

- _Location_ along the up- to downstream gradient; for long-term protection of a harbour, for example, location could be defined as distance to the harbour; for more short-term and local impacts one might consider using distance between two sediment sites; for cumulative downstream risk one might use distance of a site from the river source.

- _Energy_, i.e. potential energy [35], is also a component of the overall transport risk: the higher the energy, the higher the risk that upstream sediments be deposited downstream. The potential energy could best be represented by the slope between points of interest, or as a surrogate by the elevation difference. Some confounding factors such as barriers might interfere. The energy criteria is primarily related to the management goal of reducing sediment loads in water, and downstream, but it can also be related to the quality function.

- _Quantity_ encompasses three inter-related aspects all of which could be used as criteria: volume of contaminated sediment available for entrainment, sediment budget [36] represented by the balance between eroding and depositing material, and mobility, which describes the ease with which sediments may be entrained into flow. Mobility will be affected by the grain size distribution of sediment deposits and their composition, the length of time sediment has been deposited and the shear stress exerted on the sediment by flowing water [37]. In the Netherlands the concept of a stability constant has been used [38].

- *Quality,* which could be represented by indices of hazard, such as the example provided by Wildhaber et al. [11], or by using sediment quality guidelines (SQGs). These could be expressed as sums or means of a hazard quotient (HQ; ratios of chemical concentrations to SQGs) or biological measurements, such as acute or chronic toxicity. It should be noted, however, that for a basin-scale study, if available data are limited, screening-level chemical and biological measures can be used. Clearly, quality is linked to the goal to either control toxicity or control contaminant levels.
- *Expected risk or benefit* of a specific sediment unit to downstream sediments. This can be expressed as a combination of the likelihood of sediment moving from its current site to a site of concern and the relative quality of the two sites. At a screening level this can be expressed as the ratio of the difference in quality and the distance (horizontal or vertical) between sites.

An additional issue is time: how one selects the criteria described above depends upon the timescale of management objectives and the frequency profile of critical events in the river basin.

The scheme laid out in the above paragraphs does not account for soil erosion and soil quality issues (agricultural or industrial contaminated soils, brownfields, mine tailings, etc.). Although certainly important, in particular in a long-term perspective of sustainability [39], these issues could probably be ignored in a first evaluation. We assume that in many cases an efficient prioritisation can be achieved with a limited set of criteria; moreover, when and where erosion occurs, the sediment contamination downstream reflects soil contamination. Adding criteria accounting for these issues will therefore not necessarily increase the discriminatory power of the classification approach. Anyway, if deemed necessary supplementary criteria could be added.

3.2. A case study of the Rhine basin

The Rhine basin (figure 2) was chosen for to illustrate the concepts discussed above. It is Western Europe's largest river basin, with an area of 185000 km2, and flows through Switzerland, France, Luxembourg, Germany and the Netherlands. Many cities and major industrial areas have been located along its banks for centuries, and for decades, industrial and domestic waste flowed untreated into the river. This history is still reflected in the contamination of sediments at many sites (see [40] for details). As with many other large streams, its waters are used for various purposes, leading to conflicting priorities of interest: drinking water abstraction, energy production, ship transport up to

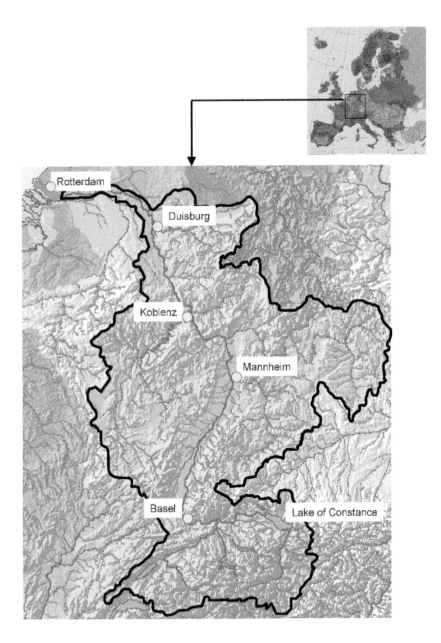

Figure 2 – The Rhine catchment area (modified on the basis of [41])

Basel, industrial and wastewater treatment plants (WWTP) releases, fisheries. Even before the adoption of the Water Framework Directive, the riparian states have expressed the ambition of restoring a good ecological condition, allowing the return of emblematic species such as salmon and sea trout. Due to the industrial past of the region, the energy production and the channelisation, complex sediment management issues have arisen. Last but not least, riparian states have set up extensive monitoring programmes, coordinated by the International Commission for the Protection of the Rhine[7]; other institutions, such as the Port of Rotterdam, also supported strategic studies, so there is quite a body of existing data on sediment issues in the Rhine basin.

3.2.1. Overview of data

The sources and nature of data collected for this case study are briefly presented hereafter.

Data availability
- Geographical data (elevation, slope, hydrography): digital elevation model (DEM) with a horizontal grid spacing of approximately 1 kilometer and a vertical accuracy of ± 30 meter, available for free at USGS (GTOPO30 system, http://edcdaac.usgs.gov/gtopo30/gtopo30.asp);
- alternatively, EuroGlobalMap (1:1.000.000) afforded by EuroGeographics (http://www.eurogeographics.org/eng/04_products_globalmap.asp).
- A more precise GIS can be obtained at the Rhine Commission (see note 7)
- Quantity: no information could be gathered on the volume of contaminated sediments at certain sites, except data on suspended matter (SPM) fluxes, which were not useable for the current exercise.
- Shear stress / mobility: the International Commission for the Protection of the Rhine (ICPR) has commissioned a study of the re-suspension risk of sediments in the Rhine. The critical shear stress has been measured in the laboratory for six different locations, i.e. upstream weirs at Marckolsheim, Gertsheim, Strasbourg, Iffezheim and Gambsheim on the Rhine, and Eddersheim on the Main. In addition, some field investigations have been done in the lower part of the catchment, at Amerongen and Hollands Diep, and at Duisburg on the Ruhr. However, quality data are not available for all these sites.
- Quality: metal total concentrations in bottom sediments top layer - means over several samples; hexachlorobenzene, and some other organic

---

[7] http://www.iksr.org/

compounds. Nevertheless, data for the latter organic compounds are rather sparse, and the whole dataset is heterogeneous.

Overall, a dataset including location, energy and quality was available for 135 sites, and a more complete dataset with the same parameters plus critical shear stress values was available for 7 sites only[8,9]. Therefore, the case study was limited to the 3 criteria: location, energy and quality.

Overview of distributions

Elevation data were obtained from the abovementioned GIS system GTOPO30; the distribution of altitudes is summarised in Table 3.

Table 3 – Number of sites according to elevation ranges

Altitude	Rhine	tributaries
0-20	34	0
20-50	5	24
50-100	3	23
100-200	14	19
200-...	6	20
Total	62	86

Location was first described as the distance to the mouth; these distances were calculated with the abovementioned GIS. The distributions of these distances are given in Table 4.

Table 4 – Number of sites according to distance to the mouth ranges

Distance	Rhine	tributaries
0-100	23	0
100-200	12	0
200-400	5	62
400-...	22	24
Total	62	86

Due to the lack of accuracy of the digital elevation model (DEM), there are some errors in both the elevation data (some are negative, other typically

---

[8] potentially more, but data difficult to gather in the time available
[9] a "site" corresponds to a location defined by its distance to a reference point in the estuary; sediment contamination at a site is usually based on a composite sample

flawed, when looking at the up- to downstream gradient) and the location data. Other errors arose because of the location data source ("river km", instead of geographic coordinates). Nevertheless, these errors will probably influence more the classification of sites according to elevation than to location.

Quality: for the Rhine itself, though many stations display "some" data, the number of stations displaying several mean concentrations is somewhat limited. The most frequently available results are for zinc, copper, lead, nickel and hexachlorobenzene. This is also the case for the tributaries, except that very few data are compiled for organics (Table 5).

Table 5: Number of sites with available sediment quality data available for the Rhine and some tributaries

	Rhine	Neckar	Sieg	Wupper	Erft	Ruhr	all_tribut.
Nb stations	62	24	7	17	15	17	80
Zn	23	0	7	15	12	15	49
Cd	4	1	0	0	12	11	24
Cu	37	24	5	16	8	16	69
Cr	16	0	6	16	0	15	37
Hg	7	1	0	8	0	0	9
Pb	17	0	7	16	13	15	51
Ni	41	18	6	17	13	16	70
PCB	7	3	0	1	0	4	8
PAH	8	1	1	8	0	1	11
HCB	51	0	0	0	0	0	0

A mean quality index based on the mean quotient formula [42] can be calculated, using the total concentrations for Cu, Ni, Pb, Zn, Cd, Cr and Hg, and hexachlorobenzene for the Rhine only. The guidelines involved in these calculations are CTT values, shown in Table 6. CTT (for chemical toxicity test) was primarily set up for making relocation decisions for dredged materials from Dutch harbours [43]. Other guidelines could be used as well, as the focus of this exercise is not on the reliability or relevance of various guidelines , but on prioritisation methods, with the proviso that any guideline applied addresses the whole range of measured compounds.

Table 6 - Guidelines used for calculating the quality index [43]

Substance	unit	CTT
Zn	mg/kg dw	365
Cd	mg/kg dw	4
Cu	mg/kg dw	60
Cr	mg/kg dw	120
Hg	mg/kg dw	1.2
Pb	mg/kg dw	110
Ni	mg/kg dw	45
Sum PCB (7 congeners)	µg/kg dw	100
Sum PAH (10 substances)	mg/kg dw	8
HCB	µg/kg dw	20

The distribution of index values within the whole dataset is presented in figure 3.

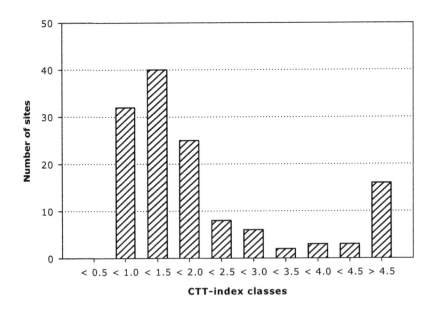

Figure 3 – Bar chart of CTT-index values

3.2.2. Application of Classification Methods

Two classification methods have been applied to prioritise sites along the Rhine basin: a score ordination and a fuzzy logic based approach. Both make use of expert knowledge but to a different extent and employ different mathematical steps. Accordingly, their acceptance by stakeholders will differ, depending on their confidence in the experts and on their knowledge of the procedures. As the stakeholders acceptance may determine eventually which approach to use for decision making, both classification methods will be applied here using the same database and the same ranking of criteria in order to examine the degree of differences in the outcomes.

In these approaches, along with the step consisting of the determination of class boundaries, ranking criteria according to their importance is critical. This step mainly uses expert judgement, nevertheless the arguments should be made clear, in order to permit further adjustments, if deemed necessary. In addition, there will be more confidence in the final classification of the criteria if it results from a consensus among experts.

On the basis of the above example, we suggest to adopt the following order:

- A- *Quality:* Any beneficial use of sediments would be impaired in case of poor quality; moreover, poor quality sediments will almost systematically involve a management decision
- B- *Mobility:* Mobile sediments are not necessarily an issue, as sediment transport is a natural process. The mobility will influence management priorities either if sediments are contaminated, or if it comes from exaggerated erosion.
- C/D- *Location*, or *Potential Energy*: Upstream sediments will move downstream more or less rapidly. So an upstream location could increase the priority level for contaminated sediments, as compared to similarly contaminated sediments located downstream. Both items are however interdependent; one of them could be sufficient, unless the relative position in the catchment could be expressed in a simple way.
- E- *Size of contaminated sediment sites:* depending of data availability.

Additional items, as explained above could include the benefits expected from a remedial action, and sediment budget.

Introduction to the Score Ordination Approach

In the score ordination approach, variables (either dependent or independent) which are considered to be essential to the studied phenomenon are selected. These variables are ordered according to their relative importance to processes of concern. As a consequence, unlike other scoring methods, the score ordination approach does not determine a specific weight for each variable, because the desired "weight" is a relative position in a list rather than an absolute value. For example, for pesticide transfer to surface waters, crop acreage is more important than the applied dose or than the solubility of the pesticide. For each variable, the magnitude of its variability is translated into three classes (e.g. H = high, M = medium, L = low) or more. Three classes afford in general a sufficient discrimination when combining all the variables. However, for variables taking a large range of values, four classes can be more appropriate. Expert judgement is needed for both the ordering of the variables and for the determination of class boundaries. The relevance of these steps is enhanced if a group of experts is involved in a consensual statement. If expert judgement is obtained by consensus from a panel involving representatives from different stakeholder groups, the classification approach and its results will be more reliable, and possibly more transparent.

With the use of a penalty table, all possible class combinations for the selected number of variables are listed and a score ranging from 1 to the maximum number of combinations is given to each possible combination. For example, if 2 variables and 3 classes for each variable are used to study a specific phenomenon then there are 2^3 (8) possible combinations of classes. Each combination is assigned a score ranging from 8 to 0. An example of such a penalty table is given below in Table 7.

Table 7 - Example of a penalty table for 2 variables, A and B using three classes, H (high), M (medium) and L (low)

Variable		Score
A	B	
H	H	8
H	M	7
H	L	6
M	H	5
M	M	4
M	L	3
L	H	2
L	M	1
L	L	0

Objects are then ranked using the "penalty table" according to their specific characteristics (values assigned to each parameters transformed into classes). This process can easily be automated with standard spreadsheet functions.

One of the main advantages of the score ordination approach is that it allows decision-makers to weight the variables without fixing a standard value for the weight. The method *per se* is neither better nor worse than another in terms of stakeholder involvement. Nevertheless, its design permits stakeholder involvement on two critical issues, namely the relative weight of each variable and the determination of appropriate boundaries between classes. Thus, score ordination performs best when used in the perspective of consensus-building, at least among a panel of experts. This can afford a better reliability of predictions, as suggested by some applications, e.g. [44].

This approach cannot be applied when some data are lacking: in that case, the item cannot be ranked, or the missing data is given the worst-case score, which can be misleading. In addition, uncertainty is not addressed; thus, some items could be improperly ranked, for one or more variables, when their supposed values are close to the class boundaries. Lastly, the relative influence of the variables on the issue at stake must be the same in all circumstances to result in appropriate rankings.

Results of the score ordination approach
According to Table 3 and Table 4, both altitude (representing the potential energy) and distance to the mouth (representing the location) can be better classified with 4 classes; according to figure 3, quality can be expressed in 3 classes, based on the median and 75th percentile of CTT-index values. These classes are reported in Table 8.

Table 8 - Class boundaries for site prioritisation in this case study

Criteria	Class 1	Class 2	Class 3	Class 4
Distance (km)	< 100	≥ 100 - < 200	≥ 200 - < 400	≥ 400
Altitude (m)	< 50	≥ 50 - < 100	≥ 100 - < 200	≥ 200 m
Quality (CTT)	< 1.4	≥ 1.7 - < 2.3	≥ 2.3	

Depending on which criterion is applied first in the score ordination approach, different penalty tables can be drawn; they all have 48 rows. For a given row, the overall score, combining individual scores for each criterion, may be expressed first as a combination of scores (e.g. 111, if the three criteria are all in class 1), and also as a rank, between 1 and 48. Two examples are given in Appendix 1: In part (a) distance to the mouth is the first criterion applied, then quality, then location (distance to the mouth), while in part (b) quality is applied first.

In this example, changing the order of criteria strongly modifies the assignment of sites in all categories. It is however more obvious for the extreme ones: there are few sites from the Rhine itself in the low-priority category (scores 1-9), and an even fewer number of sites located on tributaries in the high priority category (scores 40-48)(figure 4).

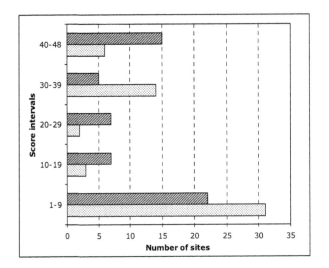

a-Rhine

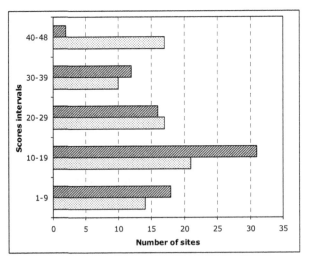

b-Tributaries

Figure 4: Variation of scores as a function of the criteria order at sites along the Rhine (a) and its tributaries (b) (*hatched bars*: quality > energy > location; *dotted bars*: energy > quality > location)

Introduction to the Fuzzy Logic Approach

The fuzzy set theory was developed by L.A. Zadeh [45] to handle highly variable, linguistic, vague and uncertain data or knowledge [46]. According to Silvert, classification of ecological impacts is complicated by uncertainty in the interpretation of complex data, the existence of gaps in data basis and the often qualitative rather than quantitative character of ecological data [27]. If compared to conventional methods based on binary logic (sharp yes-no or pass-fail decisions), fuzzy set approaches allow for a distinctly more effective management of this kind of imprecise data.

They provide a formalism for incorporating vagueness in a classification scheme. The vagueness, i.e. the uncertainty to which the gathered data are exact, to what extent they are reproducible, and whether the attributed characterisation (e.g. "no inhibition" as opposed to "moderate inhibition") is correct, is dealt with by avoiding constraints of associating each value to only one unique class. For example, strict thresholds of chemical quality classes that are based on data which can only be measured with a certain precision do not necessarily reflect the environmental quality. In a simple example, it could be assumed, that the threshold level for 10 contaminants in a sediment is 60 $\mu g.kg^{-1}$. Below this threshold, the sediment is assigned to class "medium quality" and above to class "bad quality". In conventional classification systems with strict threshold values, a sediment with 10 contaminant concentrations of around 58 $\mu g.kg^{-1}$ could accordingly be regarded as less impaired than a sediment with 9 background concentrations below 20 $\mu g.kg^{-1}$ and one contaminant of 62 $\mu g.kg^{-1}$ that slightly exceeds the limit value, although the analytical uncertainty of the chemical analyses may be far less accurate than 4 $\mu g.kg^{-1}$ sediment. In fuzzy sets (see Figure 5), uncertainty and variability are addressed as overlaps of classes. To those data that cannot be attributed to one set or the other but are in between, different degrees of membership to these fuzzy sets are assigned. Staying with the example of chemical concentrations above, the values just below the quality class threshold would already be considered as partly belonging to the next quality class. In other words, the analytical uncertainty makes it simply impossible to determine certainly to which class this sediment belongs, and the fuzzy sets theory allows to handle this ambiguous situation and nevertheless make a "good decision". The degree of overlap between two fuzzy sets depends on the uncertainty (for example the precision and accuracy of a chemical analysis) of the data that are to be classified. In the case of a precision of chemical analysis of ± 5 $\mu g.kg^{-1}$, the certainty with which a single contaminant would add to the next quality class, would gradually increase between 56 and 64 $\mu g.kg^{-1}$ (Figure 5). Instead of strict categorisations, the fuzzy approach is thus able to both create a classification characterized by a transition zone of boundaries and use the predicted gradual membership in assigning different classes during subsequent calculations. This avoids the inevitable loss of information which occurs if each variable along a decision process is forced into precise classes. Fuzzy sets allow this information to be carried to the final de-fuzzyfication step.

Another advantage of the fuzzy logic approach is that a linguistic description of an expert's classification can be used by translating it into fuzzy **if-then** rules using linguistic predicates. Fuzzy linguistic rules appear close to human reasoning and very much resemble how managers think. This helps to keep the expert system transparent. A simple fuzzy rule could e.g. be: *if quality is low AND mobility is high AND potential energy is high, then site prioritisation is high.*

Thus, the linguistic description enables the combined calculation of different measurement endpoints (e.g. different contaminant concentrations) as well as different assessment endpoints (or in this case criteria) [26].

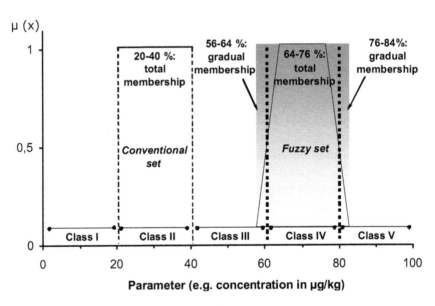

Figure 5 - Classical membership function and fuzzy set membership function (from [47, 48]
Depicted is the degree of membership function ($\mu(x)$) with reference to the different parameter values

This does not preclude the possibility of using fuzzy sets to provide precise numerical grades if these are needed for decision making. However, the process of going from fuzzy sets to quantitative indices ("de-fuzzyfication") can be carried out separately from the fuzzy analysis as a final step, minimising the truncation of information throughout.

The disadvantages of fuzzy systems, however, are that a lot of information and experience are necessary to set them up. Some of this can be resolved by the use of neural networks, but this would in turn reduce the transparency.

The fuzzy logic approach requires the development of rules on the basis of the interrelationships of the criteria. These do not necessarily have to be ranked regarding their importance as in the score ordination but can be regarded as being of equal importance. For the sake of comparison, however, the same weighting of criteria is used here as in score ordination approach.

The fuzzy rule-based system comprises the following steps:
1. Identifying the accuracy and precision of the criteria.
2. Determining overlaps between different fuzzy sets that define those areas, in which data can not be assigned to specific categories.
3. Defining the potential number of categories or classes for each criterion. If accuracy and precision of data are low, as indicated by a high variability of data, the maximum number of distinguishable categories is small.
4. Setting up the rules based on expert knowledge that will determine the prioritisation status.
5. Determining the uncertainty of the different rules, expressing e.g. lack of knowledge.

Results of the fuzzy logic approach
The same classes were chosen for the fuzzy approach that had been chosen for the score ordination. For comparison, the criteria order location>energy>quality was chosen, because location is the criterion with the largest values. Putting most emphasis on this criterion would give us the highest and most obvious differences in the outcome should there be any due to the variability of data. The following variability or uncertainties were assumed for the different criteria: Distance: ±25 km, Altitude ±50 m, Quality index: ±0.6. It has to be noted, that these uncertainties are rather large and only assumed here for the sake of the case study. For realistic calculations, the variability of each of these criteria would have to be ascertained.

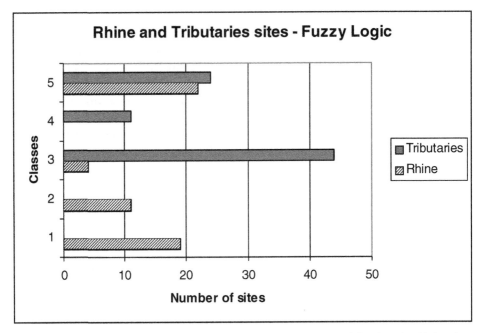

Figure 6 – Site priority according to a fuzzy logic approach (criteria weights: location higher than energy, then quality)

As can be expected by this order of criteria, the ranks of the tributaries that flow into the Rhine are determined primarily by their distance to mouth (figure 6).

In comparison to score ordination, using the same order of criteria, almost all Rhine data are classified in the same way, but the distribution of the tributaries is different (Figure 7). This is caused by the uncertainty of the location criterion which leads to a different classification result for those rivers, which distances from mouth lie between to classes. While the score ordination assigns a discrete class already to these distances, a fuzzy based system keeps this uncertainty in the calculation and the final outcome is in those cases determined by the fulfilling of the other criteria.

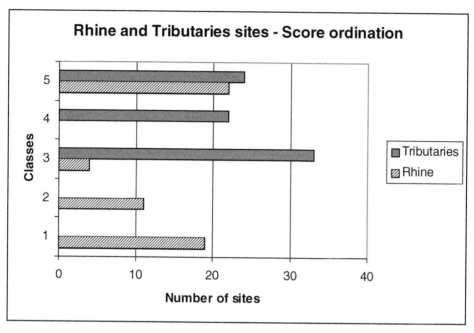

Figure 7: Site priority according to score ordination (criteria weights: location higher than energy, then quality)

3.2.3. Discussion

This case study raises various issues. First and foremost, it was driven by existing data; this is far from ideal, because each data-subset was produced in a specific context, for specific objectives. Moreover, data on important aspects, such as sediment mobility, were almost completely missing. If managers or basin authorities decided to apply such an approach in other catchments, they would unavoidably encounter the same difficulty. These difficulties could even be worse, as the Rhine dataset benefited from important and long term efforts from riparian countries within the Rhine commission. It is recommended that this issue is addressed in a dynamic perspective, that is to prioritise sites first with existing data, eventually completed by a specific study of limited ambition, and in the meantime set up a long term monitoring effort allowing to improve the site prioritisation, and revise successive basin management plans.

Several other issues are related to the prioritisation approach itself:
- *Time and spatial scales:* the case study of the Rhine river was set up in a long term/long range perspective, as if the main goal was to protect or restore the coastal areas and settlements (harbours) at the catchment's outlet. More short-term management goals should also be considered, particularly in a

water framework directive (WFD) perspective. This scaling issue is affecting both *location* and *energy* descriptors.

- In relation with this scale issue, there might also be a question regarding whether or not sites are located along an up- to downstream gradient or not, that is whether an upstream site is exerting an influence on a downstream site or not. This **connectivity** issue does not really matter when considering long term risk, or priorities from the outlet point of view. But it could be an issue in a short term perspective, or at an intermediate spatial scale. Looking at the distance between sites rather than at absolute distance to the catchment's outlet could help resolve this issue. Another way, not exclusive, could be to adopt a tiered approach.

- The various aspects of the **quantity** issue are not well addressed yet, for at least two reasons: one is the lack of data (on unit sizes, on sediment fluxes and on shear stresses), and the other is our lack of comprehension about the propagation of flows in a basin (which are the control factors and how can they be simply described).

- Another difficulty is related to the complex interplay between the prioritisation criteria. This could lead to the concept of **relative quality:** this term addresses the potential for sediments in dynamic systems to move and potentially impact other sites. Clearly, a contaminated sediment site that has a probability of spreading to less contaminated downstream sites should be managed before an equally contaminated but stable sediment site. Such a parameter is a function of the quality gradient, the energy and the sediment stability. Perhaps a function such as $(\Delta Q/\Delta L)*S$, where ΔQ is the difference in absolute quality between an upstream site and a downstream site, ΔL is the distance between sites, and S is a measure of the sediment stability, would provide a screening-level indication. Of course, if an upstream site is of better quality than downstream quality and is erosive, it has the potential of attenuating the risk of a downstream contaminated sediment, and the sediment is thus potentially beneficial. One could argue however that the most downstream sites would not be classified, and that this gradient approach allows one to examine adjacent sites 2 by 2, while even on a relatively short term basis sites are influenced by several upstream sediment units.

- Though potentially interesting, the concept of **expected benefits** [35] was not addressed in this first attempt. The main reason for that is the difficulty to find an appropriate descriptor for this concept, and accordingly to describe the related classification scale. As the potential benefits are somewhat embedded in other descriptors, e.g. quality, it could also be argued that it would be useless to increase the complexity of the prioritisation approach. This would need perhaps some more case studies.

Typically, this discussion raises serious scientific issues; it can be expected that some of them at least would require more or less complex models and large datasets, which would be in contradiction with a screening level approach. What is actually needed is to reduce the overall uncertainty associated with the prioritisation process, not to achieve a high level of precision for each parameter involved. Maybe a tiered approach could help in this context, using existing data on quality, location (distance to the catchment's outlet) and elevation (potential energy) in a first step, and then refining the classification with additional data where the associated uncertainty would be unacceptable.

As shown in the case study above, a fuzzy classification would provide results very similar to those of the score ordination approach if based on the same criteria, class boundaries, and on rules set on the same order of priority. Differences occur through the inclusion of uncertainty and variability of data. This can improve the realistic assessment of sites – if the overlaps of fuzzy sets (i.e. the uncertainty of data) are determined in a way that is acceptable to and understood by decision makers.

According to the abovementioned evaluation criteria [5], score ordination appears easy to use, robust and rather transparent. Its use is nevertheless hampered by incomplete datasets, and is less appropriate when the criteria have complex interplay. Fuzzy logic does not have necessarily a weighting of criteria and can possibly cope more easily with complex issues. In those cases, however, the large number of if-then-rules severely impedes transparency. It seems as sensitive and robust as score ordination – essentially because in fact these elements are related to the classification criteria rather than to the classification method. As both methods have their advantages and disadvantages, these should be communicated to the decision makers and their application be made dependent on the stakeholders acceptance.

4. Site specific risk assessment approaches

Basically, risk is characterized by a combination of hazard and exposure [4]. As mentioned above, in the case of sediments risk characterisation is however complicated in most instances by the co-occurrence of several *measurement endpoints* (e.g. toxicity tests) for one given *assessment endpoint* (e.g. benthic population maintenance), and even by the co-occurrence of several assessment endpoints. Moreover, toxicity tests using benthic organisms usually do not generate quantitative results, such as effective concentration for x % of individuals. So, current risk characterisation methods (risk quotient, comparison of distribution) cannot be applied. Scoring methods, because they assign values or scores to a class of measures for a particular variable, or fuzzy approaches, for the same reason, seem more appropriate in that perspective.

The estimation of uncertainty should be part of the risk characterisation step [4]; to a certain extent, WOE approaches incorporate uncertainty in the final result. By design, fuzzy logic-based approaches also cope with uncertainty.

4.1. A brief review of published approaches

A few contaminated sediment risk characterisation approaches that also deal with the uncertainty issue, are briefly reviewed below.

The Massachusetts Weight of Evidence Workgroup [49] developed a scoring system to account for several measurement endpoints within a single assessment endpoint. Several aspects have to be examined, and weighed based on the risk estimate; these aspects are split into three categories, i.e. the weight of the assessment endpoint, the magnitude of the response, and concurrence among measurement endpoints.

1. First, the weight is related to 10-11 attributes[10] grouped into three categories: relationship to the assessment endpoint, data quality and study design. A formal grid is proposed for assessing each attribute. Moreover, attributes are given a scaling value, acknowledging that some attributes are more important than others. Finally, the weight is expressed as a score, scaled from 1 to 5.
2. Second, magnitude is related to two questions which are "is there harm?", and "is it low/high?" This may involve the use of metrics, but [49] state that managers often prefer using discrete values such as low, medium or high.
3. The third component, concurrence, reflects whether measurement endpoints agree or diverge. In practice, a graphical approach is described (Figure 8), where the above mentioned components (1) and (2) constitute the axes.

Assessment Endpoint		Weighing factors (increasing confidence or weight)	
Harm	Magnitude	1 (lowest) → 5 (highest)	
Yes	High		
Yes	Low		↑
Undetermined			
No	Low		↓
No	High		

Figure 8 – Risk characterisation redrawn from [49]

---

[10] 11 attributes are cited in some parts of the paper, 10 in others

One of the main goals of this approach is to provide a structured way to include the uncertainty (here called weight) within the risk estimate. Some aspects however remain confusing (e.g. the approach to weighting is rather complex and lacking transparency), while others might need further developments or discussion (e.g., the exposure assessment, or the concept of concurrence).

In a case study of the risks caused by a former shipyard in New Hampshire, Johnston et al. [50] assigned values to five different measures of exposure and to three different measures of effect as illustrated in Table 9. The outcomes of the measures were interpreted based on whether the result added some indication of risk or not.

Table 9 - Interpretation scheme in the shipyard case study [50]

Type of measure	Degree of response	Interpretation	Value[11]
Exposure	≤ Reference condition, or below conservative benchmark concentration	Negligible exposure	0
	> Reference condition	Low exposure	1
	Statistically > reference concentration	Elevated exposure	2
	> conservative benchmark concentration	High exposure	3
	> non conservative benchmark concentration	Adverse exposure	4
Effect	Similar to reference or control condition, or below ecologically relevant threshold	No effect	0
	Worse than reference or control condition, but not statistically different	Potential effect	1
	Statistically worse than reference or control condition	Probable effect	2

These authors [50] determined risk evidence by combining the exposure and effect assessments using a matrix table as illustrated in Table 10.

---

[11] i.e. score

Table 10 - Risk evidence, according to exposure and effect assessments (from [50]

Evidence of effect	Evidence of exposure				
	0	1	2	3	4
0	0	0	1	1	2
1	0	1	2	2	3
2	1	1	2	3	3

Endpoint weights were assigned to each measure of exposure and effect, to reflect the reliability and usefulness of the measure in assessing risk to the assessment endpoint. The weighting procedure was derived from the aforementioned working group [49], and consisted of scoring the attributes of each measure as low (1), medium (2) or high (3), depending on how well the measurement data related to the assessment of stressor levels or ecological damage. Scatter plots were then used to evaluate the weight of evidence of risk for each assessment endpoint (X-axis = outcomes of exposures or effects, Y-axis = endpoint weights). Final risk for a given assessment endpoint was calculated as the average of risk scores obtained for the corresponding measurement endpoints, as shown in figure 9.

The primary advantage of this approach is that it provides a formal procedure for risk and uncertainty characterisation. The weighting process seems less complex than the approach in [49]; this may be due to the effort of standardisation of the procedure. Moreover, communication with managers may be enhanced by the use of graphical representations of exposure or effects and their respective levels of confidence. Some aspects could, however, be discussed:

- The scatter plot graphs are not very intuitive: such graphs usually represent one variable (Y) as a function of the other (X), which is not the case in these graphs.
- Too much importance is given to data quality issues. For example, Johnston et al. mention one assessment endpoint for which there was conflicting evidence between two measures which had similar weights [50]. In that case, they stated that "*no unequivocal conclusion could be identified*". Furthermore, they envisaged that "*any evidence of an effect could fix the conclusion*", and clearly rejected this reasoning, because any further measure would reduce the uncertainty. This reasoning seems to be based on the hypothesis that all measures (of effect) for a given assessment endpoint are equivalent and should therefore provide convergent results. In fact, such a similarity is not very likely: aquatic organisms used in bioassays are known to have various sensitivities, due to different exposure routes, and other

factors. In fact, when data quality is poor for a given measure, the corresponding uncertainty increases, but the risk itself is not affected [51]. If this measure is critical to a decision, it should lead to complementary investigations, rather than to a possibly incorrect conclusion.

- The exposure assessment is based on 5 classes (Table 9), but this has no actual influence on the final risk, which is based on 4 classes (Table 10).
- More importantly, exposure in this decision matrix is only represented by the sediment contamination levels: the higher the concentration, the higher the presumed exposure. However, the probability of contact between the *in situ* organisms and the contaminated media was not considered. Issues of bioavailability and mechanisms of exposure should be clearly addressed in a conceptual model leading to the decision criteria. To a certain extent, this issue can be addressed in the weighting factor that relates a given measure to an assessment endpoint, but measures of bioavailability are obviously preferable.
- Moreover, the exposure assessment should be regarded in fact as a hazard assessment, because it is based on chemical contamination levels, scaled to benchmarks or sediment quality guidelines, rather than on actual measures of direct exposure.

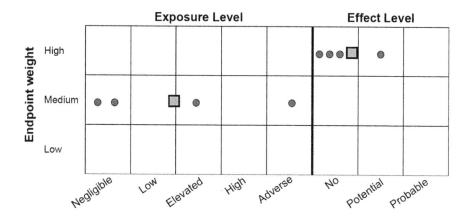

Figure 9 - Scatterplot graph of weighted measurements of exposure and effects (redrawn from [50] exemplarily for one endpoint; dots represent single measurements and squares stand for weighed averages of measurements (i.e. centroids)

Both papers provide interesting advances on several critical aspects of risk characterisation for contaminated sediments. However, two important issues still need some development, namely exposure assessment and communication.

It is thus recommended that:
4. explicit measures of exposure, related to ecological processes, should be developed;
5. the number of risk classes should be scaled to the range of possible responses, such as no further action, long-term monitoring, immediate removal and so on, provided these options may be ordered as operational responses to increasing risks;
6. risk characterisation and uncertainty evaluation should be processed separately. Both estimates should then be used by decision-makers and stakeholders.

4.2. Case study of risk characterisation for a management option

Risk assessment at the site-specific scale has been defined by SEDNET working group on 'risk management and communication' as "evaluating individual sediment parcels to determine and rank their risk relative to benchmarks, site- or basin-specific criteria." A further evaluation is needed to characterise the risk to the environment of a given management option, e.g. dredging, or to compare several options.

The application of scoring approaches similar to that described in [50] to site-specific sediment management scenarios would need:
7. to include explicit measures of exposure, related to ecological processes;
8. to scale the number of risk classes to the available management options, provided these options may be ordered as operational responses to increasing risks;
9. to process separately the risk characterisation and the uncertainty evaluation. Both estimates would then be used by decision makers (including various stakeholders).

An approach aimed at fulfilling these requirements has been proposed recently in France [52], and has been tested on a few sites [53, 54]. It applies to a specific management option, also called a scenario, for dredged material deposits in freshwater borrow pits. For the ecological part of the approach, two assessment endpoints are considered, namely "disruption of the structure and abundance of benthic invertebrate populations" and "chronic effects on organisms in the water column". The corresponding measurement endpoints involve standard bioassays on bulk sediment and pore water.

Both effects and exposure analyses follow a classification approach, based on 3 classes; thus, 2 thresholds are proposed (Table 11). In class "0", the effects are defined as similar to a reference or control condition, whilst in class "2" they are defined as significant. The thresholds were set for each measurement endpoint from the literature [38] or from existing regulations [55, 56].

Table 11 – Potential definitions of effect and exposure classes in the scenario for deposits in freshwater burrow pits (ICx %, concentration inhibiting a given endpoint, e.g. growth, for x% of the tested population; ECy %, effect concentration for x% of the population; DS: deposit surface; TS: total surface of the pit; PWV: deposit pore water volume; WCV: water column volume)

Assessment endpoint	Measurement endpoint	Class		
		0	1	2
Benthic invertebrates populations disturbance (structure / abundance)	*Effect*			
	C. riparius (survival and growth)	Mortality or growth decrease ≤10%	10%<.. ≤50%	Mortality or growth decrease >50%
	H. azteca (survival and growth)	Mortality or growth decrease ≤10%	10%<.. ≤50%	Mortality or growth decrease >50%
	Exposure			
	DS/TS: ratio between the deposit surface and the total surface	≤25%	25%<...≤50%	>50%
Chronic toxicity to aquatic organisms	*Effect*			
	Algae: P. subcapitata	Cell number x 16 (pH increase <1.5)[12]		IC50<10%
	C. dubia (survival and reproduction)	Adult females survival ≥80%, >3 broods for at least 60% females, 15 young per brood on average[12]		EC20<1%
	B. calyciflorus (survival and reproduction)	Mortality <10%; reproduction rate > 4		EC20<1%
	Exposure			
	PWV/WCV: ratio between pore water and water column volumes	≤0.1%	0.1%<...≤1%	>1%

---

[12] limit conditions in the control

Although the actual stressor, in most cases, would be made of toxic substance mixtures, the exposure classes refer to the exposure pathways, that is to the media to which the benthic or aquatic species are exposed. Thus, for the benthos assessment endpoint, the classes of exposure refer to the ratio (DS/TS) of the surface covered by the deposit (DS) to the total surface of the quarry (TS). For the water column assessment endpoint, these classes refer to the ratio (PWV/WCV) of the pore water volume (PWV) to the total volume of water in the quarry (WCV). Both volumes are estimated, the former according to the percentage of water in the bulk sediment, the latter from the dimensions of the quarry. As no reference was found in literature, the thresholds for the DS/TS ratio were set at 25% and 50%. The threshold of 1% for class "2" of the aquatic organism assessment endpoint was chosen because a risk quotient of 1 or above is expected when this level of exposure is combined with a level of effect also in class "2".

The evidence of risk is based upon the combination of the outcomes of effects and exposure assessments; the definitions of risk are summarized in Table 12. The rationale for building this table was as follows: risk classes, which can be expressed as numbers (0, 1, 2) as well as qualitative adjectives (negligible, intermediary, high), were considered as equivalent to scores, because it was necessary to find a way to combine individual risks to benthic organisms into an overall risk to benthos. Furthermore, priority was given to neither exposure nor effect; in other words, the risk resulting from the combination of effects in class 1 and exposure in class 2 was considered as equivalent to the risk resulting from effects in class 2 and exposure in class 1.

Table 12 – A possible definition of risk classes

Effect class	Exposure class		
	0	1	2
0	Negligible	Negligible	Intermediary
1	Negligible	Intermediary	High
2	Intermediary	High	High

A graphical example showing the influence of exposure (here deposit volume) on the risk scores for the benthos assessment endpoint at 2 different sites is given in Figure 10. The evidence of risk as summarized in Table 12 is based upon the sum of the respective scores for effects and exposure. This rather simple scheme leads to the ranking of risk on a constant basis: the same class will be assigned to a given risk score, whatever the combination of exposure and effects. High risk is obtained either by the combination of moderate effects and high exposure, or moderate exposure and strong effects. However, two of the combinations displayed in Table 12 may appear problematic, namely effect in class 0 and exposure in class 2, or effect in class 2 and exposure in class 0. For effects on pelagic species, a score of 0 means that the test would yield results not statistically distinct from the control, while for chironomids it corresponds to conditions close to or stricter than control conditions in OECD guidelines [38] or in the current French standard (AFNOR NF T 90-338-1). Thus, classifying these samples in a class of intermediary risk would be a cautionary decision, which can be justified either by the generic uncertainty associated with toxicity testing, or by the fact that exposure classes were arbitrarily defined. Furthermore, the same cautionary approach should prevail when effects are in class 2 and exposure in class 0, as the exposure class limits were set arbitrarily. These two combinations should not be considered as problematic *per se*; the main difficulty here is rather to establish relevant exposure class limits, which now depend upon field experiments aimed at validating this approach.

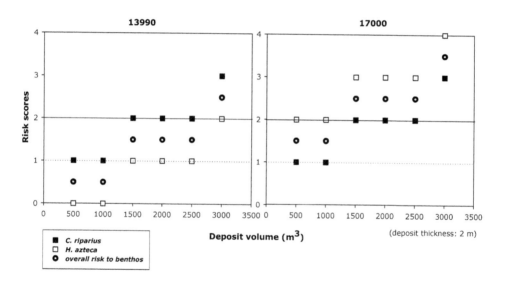

Figure 10 – Risk scores for benthos as a function of exposure (from [53]

The uncertainty evaluation is based upon modifications of the approaches of [49] and [50]. Among the known sources of uncertainty in ecological risk assessment [4], those related to the functioning of ecosystems, or to the design of the conceptual model [57] will have a nearly constant impact on the overall uncertainty if the conceptual model and the measurement endpoints are fixed, which is the case here because the framework under development is intended as guidance for local managers.

Four criteria were proposed, for the exposure assessment (1) the strength of association between the measurement endpoint and assessment endpoint, (2) the spatial representativeness, (3) the temporal representativeness, and (4) the use of a standard method. For the effect assessment, the four proposed criteria are (1) the strength of association between the measurement endpoint and assessment endpoint, (2) the distinction between effect and no effect, (3) the sensitivity, and (4) the use of standard methods. The uncertainty for each measurement endpoint is then compiled following a score ordination approach. Then, the arithmetic mean of these scores (or ranks) is calculated for each assessment endpoint.

In this context, score ordination is quite easy to handle, and fits well with most of the requirements listed by Burton et al. [5] for WOE: viz. robustness, methodology, sensitivity, appropriateness and transparency. Although proposed for WOE, these criteria can have a broader application. The robustness of the score ordination approach could eventually be discussed, as this item refers to "the consistency in interpretation irrespective of when and where conducted". In the proposed context however, inconsistencies would probably come rather from the attribute scoring step than from the score ordination approach itself. Methodology refers to the ease of use, which seems obvious in this case. With 81 possible combinations for 4 attributes, this approach is not only sensitive and appropriate according to Burton et al. [5], i.e. applicable for a wide range of conditions, it also affords some flexibility in the framework design: some measurement endpoints could be added without changing the uncertainty assessment frame.

The application of fuzzy logic based risk assessment would again need to start with the evaluation of the variabilities and accuracy of the assessment endpoints. Reproducibility of controls, of positive controls, and of replicates can be used to determine, what certainty can be attributed to different ecotoxicological values [58]. Similarly, overlaps of fuzzy sets for the measurement endpoints for exposure would need to be determined, based on the natural and methodological accuracy and precision with which exposure can be quantified. Risk classes would be calculated from the table 12, combining effect classes and exposure classes by if-then-rules.

5. Conclusion and recommendations

As is the case of other environmental issues, sediment management involves different spatial scales, and appropriate methods for selecting either the management priorities, at catchment scales, or the right management options, at a more local scale, where finally concrete actions (dredging, remediation, source control etc.) will occur. *Site prioritisation* is the process providing a site screening assessment on the basis of a pre-defined set of criteria. In a further step, risk *ranking* will allow to refine the risk appraisal on a site-specific basis and select the optimal management options. Both processes rely upon the basic principles of ecological risk assessment [4].

A prerequisite for sediment management at the river basin scale is the integration of site prioritisation and site-specific assessment schemes. This applies particularly

for site prioritisation, at basin scale:

- The prioritisation criteria should account for sediment quality, quantity, potential energy and/or mobility, and location within the catchment.
- Some of these criteria, in particular those related to quantity and mobility, are not accessible to simple descriptors and metrics. Some specific developments are therefore needed.
- Moreover, whatever the criterion, the key issue is related to the ability to describe the respective gradients along the stream and its tributaries. A function involving the quality gradient, the distance between sites and a measure of sediment stability could help integrating these key criteria.
- A gradual (tiered) approach to prioritising sites might also be helpful, and would in particular help to handle uncertainties more correctly.

for risk assessment / risk characterisation at local scale:

- The development of explicit measures of exposure, related to ecological processes, which must be carried out considering potential management options (or scenarios).
- The harmonization of risk assessment approaches, as addressed in chapter 3.
- Monitoring after sediment management, both of the treated sites, and the disposal site if this kind of management is used: data should be collated, in order to "validate" effect or exposure assumptions. These data will in turn help to improve the risk assessment frameworks and risk characterisation approaches.

Various classification approaches may be used for site prioritisation and risk ranking purposes; two of them were more discussed in detail, that is score ordination and fuzzy logic. Both have advantages, and can be recommended for site prioritisation purposes. Fuzzy logic can also be used for risk ranking at local scale, while score ordination could still be useful for some aspects of the process at this scale, e.g. uncertainty assessment. Score ordination appears easy to use, robust and rather transparent. Its use is nevertheless hampered by incomplete datasets, and probably less appropriate when the criteria have complex interplay. Fuzzy logic would probably help to cope more easily with complex issues, and seems as robust as score ordination. Ease of use, and communication would most probably be more challenging. The results from both approaches are valuable for sediment managers responsible for the allocation of resources for remediation efforts as these approaches identify specific sites which represent the greatest screening risk and this should be prioritized. As such, both approaches can also be helpful for handling the risk perception issue, by involving stakeholders in their tuning to the specific basin characteristics (setting class boundaries, and setting priority rules).

Management, and prioritisation accordingly, should be viewed as a dynamic process: "river basin management plans" need to be revised over time, so as the prioritisation process itself, which would benefit from complementary monitoring data. Geo-referenced databases including information on quality and quantity should be available at basin scale; these databases would indeed be very helpful for improving the prioritisation process design. Links to datasets generated on a site-specific basis for risk ranking purposes would also be very helpful.

References

1. Apitz SE, Power EA (2002): From risk assessment to sediment management. An international perspective. J. Soils and Sediments 2(2): 1-6.
2. Power M, McCarty LS (1998): A comparative analysis of environmental risk assessment/risk management frameworks. Environmental Science and Technology 32: 224A-231A.
3. Suter B, Vermeire G, Munns W, Sekizawa J (2003): Framework for the integration of health and ecological risk assessment. Human and Ecological Risk Assessment 9(1): 281-301.
4. USEPA (1998) Guidelines for Ecological Risk Assessment. U.S. Environmental Protection Agency: Washington D.C., USA. 159 pp.
5. Burton GA, Chapman PM, Smith EP (2002): Weight-of-evidence Approaches for Assessing Ecosystem Impairment. Human and Ecological Risk Assessment 8(7): 1657-1673.
6. Burton GA, Batley GE, Chapman PM, Forbes VE, Smith EP, Reynoldson T, Schlekat CE , den Besten PJ, Bailer AJ, Green AS, Dwyer RL (2002): A Weight-of Evidence Framework for Assessing Sediment (Or Other) Contamination: Improving Certainty in the Decision-Making Process. Human and Ecological Risk Assessment 8:1675-1696.
7. Hokkanen J, Salminen P (1997). ELECTRE III and IV Decision Aids in an Environmental Problem. Journal of Multi-Criteria Decision Analysis 6(4): 215-226.
8. Roy B (1968). Classement et choix en présence de points de vue multiples (la méthode ELECTRE). RIRO 8: 57-75.
9. Chapman PM, McDonald BG, Lawrence GS (2002): Weight-of-evidence Issues and Frameworks for Sediment Quality (and other) Assessments. Human and Ecological Risk Assessment 8(7): 1489-1515.
10. Grapentine L, Marvin C, Painter S (2002): Initial Development and Evaluation of a Sediment Quality Index for the Great Lakes Region. Human and Ecological Risk Assessment 8(7): 1549-1567.

11. Wildhaber ML, Schmitt CJ (1996): Hazard ranking of contaminated sediments based on chemical analysis, laboratory toxicity tests, and benthic community composition: prioritizing sites for remedial action. Journal of Great Lakes Research 22(3): 639-652.

12. Chapman PM (1990). The sediment quality triad approach to determining pollution-induced degradation. The Science of The Total Environment 97-98: 815-825.

13. Chapman PM (2000): The Sediment Quality Triad: then, now and tomorrow. International Journal of Environment and Pollution 13(1-6): 351-356.

14. Green RH, Montagna P (1996): Implications for monitoring: Study designs and interpretation of results. Canadian Journal Of Fisheries And Aquatic Sciences 53: 2629-2636

15. DelValls TA, Forja JM, Gonza´lez-Mazo E, Blasco J, Go´mez-Parra A (1998): Determining contamination sources in marine sediments using multivariate analysis. Trends Anal Chem 17: 181–192.

16. DelValls TA, Forja JM, GomezParra A (1998): The use of multivariate analysis to link sediment contamination and toxicity data to establish sediment quality guidelines: An example in the Gulf of Cadiz (Spain). Ciencias Marinas 24(2): 127-154.

17. Shin PKS, Fong KYS (1999): Mutiple discriminant analysis of marine sediment data. Marine Pollution Bulletin 39: 285-294.

18. Chapman PM, Hollert H (2006): Should the sediment quality triad become a tetrad, a pentad, or possibly even a hexad? Journal Of Soils And Sediments 6(1): 4-8.

19. Canfield TJ, Kemble NE, Brumbaugh WG, Dwyer FJ, Ingersoll CG, Fairchild JF (1994): Use of benthic invertebrate community structure and the sediment quality triad to evaluate metal-contaminated sediment in the Upper Clark Fork River, Montana. Environ Toxicol Chem 13: 1999–2012.

20. Aurousseau P, Gascuel-Odoux C, Squividant H (1998): Eléments pour une méthode d'évaluation du risque parcellaire de contamination des eaux superficielles par les pesticides. Etude & Gestion des Sols 5(3): 143-156.

21. Babut M, Breuzin C (2000): Pertinence d'une sélection des pesticides à mesurer dans un réseau de surveillance de la qualité des eaux superficielles à l'aide d'une méthode simplifiée d'évaluation des risques. Revue des Sciences de l'Eau, 34: 385-392.

22. Vaillant M, Jouany JM, Devillers J (1995): A multicriteria estimation of the environmental risk of chemicals with the SIRIS method. Toxicology Modeling 1(1): 57-72.

23. Guerbet M, Jouany JM (2001): Value of the SIRIS method for the classification of a series of 90 chemicals according to risk for the aquatic environment. Environmental Impact Assessment Review 22: 377-391.

24. Chabot R (2003): Etudes de suivi des effets sur l'environnement des fabriques de pâtes et papier: synthèse des 48 études réalisées au Québec dans le cadre du cycle 2. Environnement Canada, Direction de la Protection de l'Environnement: Montréal (Québec, Canada).

25. Kruse R (1996): Fuzzy-Systeme - Positive Aspekte der Unvollkommenheit. Informatik-Spektrum 19: 4-11.

26. Bojorquez-Tapia LA, Juarez L, Cruz-Bello G (2002): Environmental Assessment. Integrating Fuzzy Logic, Optimization, and GIS for Ecological Impact Assessments. Environmental Management 30(3): 418-433.

27. Silvert W (1997): Ecological Impact Classification with Fuzzy Sets. Ecological Modelling 96: 1-10.

28. Hendricks-Franssen HJWM, Eijnsbergen ACv, Stein A (1997): Use of spatial prediction techniques and fuzzy classification for mapping soil pollutants. Geoderma 77: 243-262.

29. McBratney AB, Odeh IOA (1997): Application of fuzzy sets in soil science: fuzzy logic, fuzzy measurements and fuzzy decisions. Geoderma 77: 85-113.

30. Regan HM, Colyvan M, Burgman MA (2000): A proposal for fuzzy International Union for the Conservation of Nature (IUCN) categories and criteria. Biological Conservation 92(1): 101-108.
31. Mohamed, A.M.O. and K. Côté, Decision analysis of polluted sites - a fuzzy set approach. Waste Management, 1999. 19: 519-533.
32. Tran LT, Knight CG, o'Neill Rv, Smith ER, Ritters KH, Wickham J (2002): Environmental Assessment: Fuzzy decision analysis for integrated environmental vulnerability assessment of the Mid-Atlantic Region. Environmental Management 29(6): 845-859
33. Heise S, Maaß V, Gratzer H, Ahlf W (2000): Ecotoxicological Sediment Classification - Capabilities and Potentials - Presented for Elbe River Sediments. BfG- Mitteilungen Nr. 22 - Sediment Assessment in European River Basins: 96-104
34. Hollert H, Dürr M, Erdinger L, Braunbeck T (2000): Cytotoxicity of settling particulate matter and sediments fo the Neckar river (Germany) during a winter flood. Environmental Toxicology and Chemistry 19(3): 528-534
35. Apitz SE, White S (2003): A conceptual framework for river-basin-scale sediment management. J Soils & Sediments 3(3): 125-220
36. Kern U, Westrich B (1997): Sediment Budget Analysis for River Reservoirs. Water, Air, & Soil Pollution 99(1-4): 105-112.
37. Haag I, Kern U, Westrich B (2001): Erosion investigation and sediment quality measurements for a comprehensive risk assessment of contaminated aquatic sediments. The Science of The Total Environment 266 (1-3): 249-257.
38. Den Besten PJ, Schmidt CA, Ohm M, Ruys MM, Van Berghem JW, Van de Guchte C (1995) Sediment quality assessment in the delta of rivers Rhine and Meuse based on field observations, bioassays and food chain implications. Journal of Aquatic Ecosystem Health 4: 257-270.
39. Salomons W, Brils J (Eds.). 2004. Contaminated Sediments in European River Basins. European Sediment Research Network SedNet. EC Contract No. EVKI-CT-2001-20002, Key Action 1.4.1 Abatement of Water Pollution from Contaminated Land, Landfills and Sediments. TNO Den Helder/The Netherlands
40. Heise S, Förstner U, Westrich B, Jancke T, Karnahl J, Salomons W (2004): Inventory of Historical Contaminated Sediment in Rhine Basin and its Tributaries. on behalf of the Port of Rotterdam. October 2004. 225 pp.
41. Vogt J, Colombo R, Paracchini ML, Jager Ad, Soille P (2003). CCM River and Catchment Database for Europe, Version 1.0 ed. EC-JRC
42. MacDonald DD, Ingersoll CG, Berger TA (2000): Development and evaluation of consensus-based sediment quality guidelines for freshwater ecosystems. Archives of Environmental Contamination and Toxicology 39(1): 20-31.
43. Stronkhorst J, Schipper CA, Honkoop J, van Essen K (2001): Disposal of dredged material in Dutch coastal waters; a new effect-oriented assessment framework. National Institute for Coastal and Marine Management/ RIKZ, The Hague. Report RIKZ/2001.030.
44. Babut M, Perrodin Y, Bray M, Clément B, Delolme C, Devaux A, Durrieu C, Garric J, Vollat B, Bécart D, Charrier C (2002): Évaluation des risques écologiques causés par des matériaux de dragage : proposition d'une approche adaptée aux dépôts en gravière en eau. Revue des Sciences de l'Eau, 15 (3): 615-639.
45. Zadeh LA (1965): Fuzzy sets. Information and Control 8: 338-353.
46. Adriaenssens V, Baets BD, Goethals PLM, Pauw ND (2004): Fuzzy rule-based models for decision support in ecosystem management. The Science of the Total Environment 319: 1-12
47. Hollert H (2001): A combined approach for the evaluation and assessment of the ecotoxicological load of aquatic sediments and suspended particulate matter, in Fakultät für Biowissenschaften. Ruprecht-Karls-Universität: Heidelberg.

48. Hollert H, Heise S, Pudenz S, Brüggemann R, Ahlf W, Braunbeck T (2002): Application of a Sediment Quality Triad and different statistical approaches (Hasse Diagrams and Fuzzy Logic) for the comparative evaluation of small streams. Ecotoxicology 11: 311-321
49. Menzie C, Henning MH, Cura J, Finkelstein K, Gentile J, Maughan J, Mitchell D, Petron S, Potocki B, Svirsky S, Tyler P (1996): Special report of the Massachussetts Weight-of-Evidence Workgroup: a Weight-of-Evidence approach for evaluating ecological risks. Human and Ecological Risk Assessment 2(2): 277-304.
50. Johnston RK, Munns (Jr) WR, Tyler PL, Marajh-Whittemore P, Finkelstein K, Munney K, Short FT, Melville A, Hahn SP (2002): Weighing the evidence of ecological risk from chemical contamination in the estuarine environment adjacent to the Portsmouth naval shipyard, Kittery, Maine, USA. Environmental Toxicology and Chemistry 21(1): 182-194.
51. Germano JD (1999): Ecology, statistics, and the art of misdiagnosis: the need for a paradigm shift. Environmental Reviews 7: 167-190.
52. Babut M, Perrodin Y, Bedell J-P, Clement B, Cosnier S, Corriger B, Delmas H, Delolme C, Devaux A, Miege C, Pery A, Roulier J-L, Vollat B (2003): Méthodologie d'évaluation écotoxicologique de matériaux de dragage: tests de la démarche et essais d'optimisation. 2004, CETMEF, VNF. 100 pp.
53. Babut M, Delmas H, Bray M, Durrieu C, Perrodin Y, Garric J (in press): Characterizing the risks to aquatic ecosystems: a tentative approach in the context of freshwater dredged material disposal. Integrated Environmental Assessment and Management.
54. Bray M, Babut M, Vollat B, Montuelle B, Devaux A, Bedell JP, Delolme C, Durrieu C, Clément B, Perrodin Y, Triffault G (2004): Evaluation écotoxicologique de matériaux de dragage: Application à 5 sédiments du Nord-Pas de Calais. 2004, CETMEF,VNF. 147 pp
55. MATE, Décret 2002-540 du 18 avril 2002 relatif à la classification des déchets - NOR: ATEP0190045D, M.d.l.A.d.T.e.d. l'Environnement, Editor. 2002: J.O. n° 93 du 20 avril 2002. 7074.
56. MATE, Critères et méthodes d'évaluation de l'écotoxicité des déchets / Criteria and evaluation methods of the ecotoxicity of wastes, M.d.l.A.d.T.e.d. l'Environnement, Editor. 1997. 35 pp.
57. Vorhees DJ, Kane Driscoll SB, von Stackelberg K, Cura JJ, Bridges TS (2002): An evaluation of sources of uncertainty in a dredged material assessment. Human and Ecological Risk Assessment 8(2): 369-389.
58. Ahlf W, Heise S (2005): Sediment toxicity assessment: Rationale for effect classes. JSS - J Soils & Sediments 5(1): 16-20

Appendix 1   Prioritisation at river basin scale, risk assessment at site-specific scale: suggested approaches

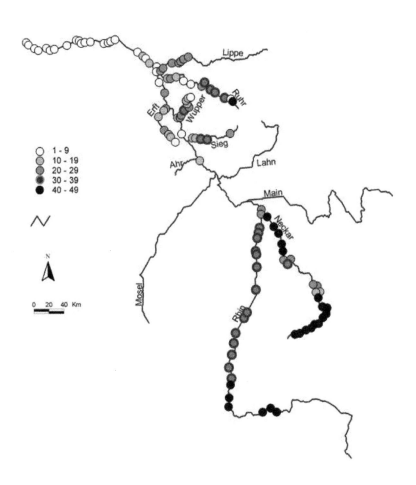

(a) Site prioritisation by score ordination

Three criteria used, in the following order: location, i.e. distance to the mouth, sediment quality, and energy i.e. altitude

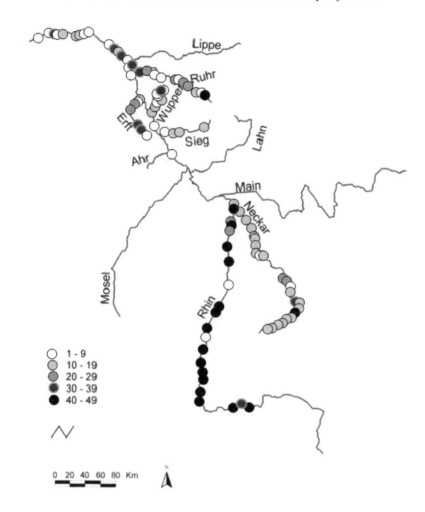

(b) Site prioritisation by score ordination

Three criteria used, in the following order: sediment quality, energy i.e. altitude, and location, i.e. distance to the mouth.

Sustainable Management of Sediment Resources: Sediment Risk Management and Communication
Edited by Susanne Heise
153

Risk assessment approaches in European countries

Pieter J. den Besten [1]

[1] *Institute for Inland Water Management and Waste Water Treatment (RIZA), Ministry of Transport, Public Works and Water Management (RIZA), P.O. Box 17, 8200 AA Lelystad, The Netherlands*
E-mail: p.dbesten@riza.rws.minvenw.nl; see *also www.akwka.info*

1. Introduction

Risk assessment is defined as the procedure in which the risks posed by inherent hazards involved in processes or situations are estimated either quantitatively or qualitatively (EEA: http://service.eea.eu.int/envirowindows/chapter1h.shtml). Risk itself is defined as expected losses (of lives, persons injured, property damaged and economic activity disrupted) due to a particular hazard for a given area and reference period. The concept and analytical techniques for risk assessment were initially developed for technological systems (space technology, nuclear installations) and were later adapted to assess risk to human health [1,2]. During the last two decades, the application of risk assessment for the evaluation of threats to the environment has gained interest.

With respect to sediment management, the risks related to sediment may deal with quantitative and qualitative aspects (see also *Chapter 2 - Sediment Management Objectives and Risk Indicators*). Assessment of the first may require calculations of the economic and personal costs of processes such as the loss of navigation, flooding events and of erosion of fertile soils. The second type of risk relates to the ecological, biological and chemical quality of sediments in our rivers, which in many cases has been impaired by human activities. Risk assessment is the step where sufficient information should be gathered to enable the responsible authorities ("risk managers") to 1) determine whether risks exist, and if so, 2) formulate management options to reduce risks to acceptable levels. A crucial element of risk assessment is the application of criteria that help risk assessors and decision makers to determine if action is needed at all.

The management goals that require a risk assessment can be manifold: finding an acceptable destination for dredged material (e.g. prediction of risks after relocating dredged material on land, including consequences for crops and livestock), sediment remediation, an evaluation of risks caused by sedimentation during flooding, etc. (see Chapter 2). This chapter will deal with one particular but important qualitative aspect of risks related to sediments: the risks caused by contaminants present in the sediments. Europe, like other densely populated areas around the world, has a problem of historic contamination of sediments. In line with the Water Framework Directive, this urges the development of integrated sediment management on a river-basin scale. Risk management is necessary for the improvement of the ecological status of water bodies, but also to reduce risks for human health and to protect groundwater resources. The basic elements of risk assessment are covered comprehensively by e.g. Van Leeuwen and Hermens [3] or US EPA [4], or can be found at EEA and US EPA internet sites. It is important to realize that risk assessment is first of all a process, in which risk assessors, preferably together with stakeholders and risk managers, follow the chain of problem formulation, exposure & effect assessment, risk characterization and communication, which should eventually lead to adequate risk management. In Europe sediment risk assessment mostly deals with the ecological risks of contaminated sediments. For this reason most of the information and discussion in the present chapter is about ecological risk assessment. However often it is not realized that sediment contamination can also have negative consequences for human health and for the quality of groundwater and surface water. Therefore also risk assessment approaches for these aspects are described in the present chapter.

1.1. Ecological risk assessment

A definition of risk related more closely to ecotoxicology (the science describing fate and effects of increased levels of elements (e.g. heavy metals) and of xenobiotics is: "the ecological risk of a certain substance is the product of the hazard (for the species of a particular ecosystem) and the probability that exposure occurs". The hazard of the substance depends in this definition on specific chemical properties of the substance (i.e. its toxicity) and on the sensitivity of certain species that are regarded as representatives of the ecosystem. Because it is not possible to study all species of an ecosystem, risk assessment approaches will have to focus on a selection of species. Which particular species are chosen should depend on taxonomic and ecological information (selection of different taxonomic groups, ecological strategy, place in food web etc) and knowledge of the relative sensitivity for toxic effects in comparison to other species. Because such information is often not available, assessment of the ecological risks is carried out in multi-tiered approaches, intended to increase the weight of evidence step by step as more proof is found for the vulnerability of essential ecosystem components [5]. It has also been a major reason for the development of bioassays, experiments in which well-defined test species are used in the laboratory as surrogate species for assessing the ecological risk.

The term 'exposure' in the definition of risk illustrates the important role of assessing the actual exposure condition. An important aspect in the risks caused by sediment-bound chemicals is the degree (and mode) of exposure encountered by sediment-dwelling organisms. As a result of sorption of contaminants to sediment particles, actual exposure levels are often lower than would be expected on the basis of the total concentrations of compounds in sediment [6,7]. However, it is also known that mixtures of contaminants can have additive or synergistic effects [8,9], which may not be well addressed by single chemical sediment quality guidelines (SQGs). Even the identity of an often significant portion of the contaminants is not known. For these reasons, ecological risk assessment and sediment quality assessment have been based not only on chemical measurements, but also on biological endpoints. Biological endpoints integrate the effects of all contaminants present at their actual bioavailability (and detect possible combination or synergistic effects). Finally, ecological risk assessment of contaminated sediments should also account for possible exposure through diet [10,11].

Most so-called ecological risk assessment (ERA) frameworks deploy biological effects-based sediment quality assessments. ERA frameworks may differ greatly with regard to their focus points. The focus may be on effect modeling and the use of chemical standards (sediment quality criteria derived from data on laboratory tests or field surveys). Simple risk assessment can be based on calculation of PEC/NEC ratios (PEC = predicted environmental concentration; NEC = no effect concentration). More advanced approaches deploy effect modeling. An example of this is the calculation of the potentially affected fraction of species (PAF; [12]). Other ERA frameworks focus on sediment quality assessment by bioassays performed in the laboratory or in the field (*in situ* bioassays) (see [13]). Finally, the ERA can be based primarily on field observations, e.g. on macrozoobenthos or on bird populations (see e.g. [14,15]). It is not always easy to link effects (e.g., reduction in breeding success of fish-eating birds) to sediment quality as the cause. Because biomarkers can signal very contaminant (group)-specific responses in organisms from the field, there can be an important role for biomarker techniques in field surveys [16]. The triad approach has proven to be a successful combination of the three categories of measurements: chemical analyses, laboratory tests and field observations [17]. In the triad approach, bioaccumulation levels can also be used as field observations [18]. In relation to ERA frameworks, the term 'bioassay' is used more frequently than 'toxicity test' or 'biotest'. While 'biotest' can be regarded as a more general term, some risk assessors use the term 'bioassay' for a test with sediment, sediment pore water, or other field samples, while the term 'toxicity test' is used for tests with selected chemical(s) dosed by the investigator.

The present paper is intended to give an overview of the use of ecological risk assessment frameworks in Europe. For an overview of the use of sediment quality guidelines in Europe the reader is also referred to a paper by Babut *et al.* [19]. One of the important points that need to be addressed is that in this paper the term 'decision' will mean different things for different countries. The information available will vary, as will the depth of environmental policies. Therefore, it is important to know the purpose of ERA in the different countries and to relate that to their sediment management programs. Two management goals for ERA are directly related to the EU water framework directive (WFD):

- Assessment of the ecological risks caused by contaminated sediments (henceforth referred to as '*in situ* sediment risk assessment')

- Assessment of risks of contaminants in sediment after dredging (henceforth referred to as 'risk assessment for dredged material management' or '*ex situ* sediment risk assessment').

These two risk assessment approaches will be outlined below.

In situ sediment risk assessment

In situ sediment risk assessment deals with the deleterious effects of historical pollution on certain parts of the ecosystem (i.e. the benthic community). However, *senso stricto* for the EU WFD, the possible negative effect of sediment contamination on the ecological quality of the water is the key issue. Sediment remediation (also called environmental cleanup) in combination with source control could be the preferred management option to solve this problem. Before such a decision could be made, the risk manager should have a clear picture of what the problem is. Therefore studies should be initiated that focus on the 'direct' ecological risks of sediment contamination or 'indirect' risks. Direct ecological risks refer to contaminants causing effects on sediment-dwelling organisms, e.g. with negative consequences for biodiversity. Indirect risks include two mechanisms: 1) substances bioaccumulate in aquatic food chains leading to effects on higher trophic levels or 2) decreased biomass production in the sediment as a result of the afore-mentioned direct effects limits food supplies for higher trophic levels. In addition, the risks for aquatic species related to recontamination of the surface water should be assessed. Part of the assessment is a detailed physical and chemical characterization of the sediment, preferably including measurements of the bioavailable fraction of contaminants in the sediment. The direct effects are usually assessed (in multi-tiered approaches) by performing inventories of the benthic zoomacro-invertebrate community and by measuring sediment toxicity in laboratory or field bioassays. The assessment of possible indirect effects is mostly limited to an assessment of potential effects occurring through food chain poisoning. This may or may not account for effects occurring via recontamination of the surface water. *In situ* sediment risk/quality assessment focuses on location-specific conditions with respect to the bioavailability of contaminants and the assessment of the damage to the ecosystem, which tends to be retrospective, but some elements such as assessment of food chain poisoning and recontamination of surface water can be regarded as prognostic.

Risk assessment for DM management

The second management goal closely related to the WFD is the management of dredged material (DM). Here, the assessment should produce data that help the authorities to choose between management options for the relocation, disposal (aquatic or on land), or treatment of dredged material. Multi-tiered risk assessment approaches are also commonly used for this purpose. As with *in situ* sediment risk assessment, a detailed physical and chemical characterization of the sediment is carried out. The use of measurements of the bioavailable fraction of contaminants in the sediment is only relevant when the future disposal or re-use conditions can be mimicked. Bioassays are often included in the sediment quality assessment or added as a second tier. The approach is more prognostic, i.e. based on the outcome of the assessment, and predictions are made of the consequences of free disposal of dredged sediments in the environment. In that respect, this approach using sediment toxicity assessment resembles total effluent risk assessments (see e.g. [20,21]).

Comparing the two different risk assessment approaches, it is likely that they lie at very different levels in a decision-making process. *In situ* sediment risk assessment is usually a front-end investigation necessary to evaluate whether sediments are a risk, before any decision about an action would be needed. Risk assessment for dredged material management is something that is carried out after dredging has already been proposed (e.g. dredging for nautical reasons), but when disposal options have to be considered [22].

1.2. Human health risk assessment

Apart from ecological risks, other risks may also require attention in the *in situ* sediment risk assessment, such as possible negative effects on human health and transport of contaminants to deeper sediment layers and subsequently to the ground water. Effects on human health can arise from exposure to contaminants through water (e.g. during swimming, through inhalation of chemicals volatilizing from sediment or through groundwater contamination, see below), fish consumption (freshwater fish or sea food), or in the case of small children, by hand–mouth contact while playing in shallow waters. A number of models can be applied to estimate the exposure of humans to contaminants from water, sediment and fish.

Human health risk assessment (see [1,3,23]) involves hazard identification, toxicity assessment, exposure assessment, and risk characterization. Both cancer and non-cancer health effects for adults and children are evaluated. Carcinogens are evaluated based on the weight of evidence and potency. The weight of evidence concept qualitatively assesses whether a chemical is known to cause cancer in humans, likely to cause cancer in humans based on animal data and limited human data, or not likely to cause cancer in humans. The chemical-specific potency is based on the cancer slope factor – a plausible upper-bound estimate of the probability of a response per unit intake for a chemical over a lifetime. The slope factor combined with exposure information is used to estimate an upper-bound probability of an individual developing cancer as a result of exposure to a particular level of a potential carcinogen over a lifetime.

Non-cancer health effects are evaluated using a Reference Dose (RfD) for oral and Reference Concentration (RfC) for inhalation. The RfD and RfC are defined as an estimate (with uncertainty spanning perhaps one order of magnitude or greater) of a daily exposure level for the human population, including sensitive sub-populations that are likely to be without an appreciable risk of deleterious effects during a lifetime. Comparison of the exposure dose over a specific time frame to the RfD indicates a concern for potential non-cancer health effects.

At sites with sediment contamination, routes of exposure may include ingestion of contaminated river water, inhalation of chemicals volatilizing from sediment, recreational exposures (incidental ingestion of sediment and dermal contact with sediment and water), and consumption of fish (or seafood in general). Usually the primary risk is from ingestion of fish (or mussels, crustaceans) where chemical-specific concentrations from sediment bioaccumulate. Exposure from fish consumption involves determining the fish chemical concentration, the daily amount of fish ingested, the frequency of fish obtained from a specific source, and the duration of exposure.

1.3. Groundwater contamination

Groundwater contamination can be the result of transport of contaminants from surface water or sediment to deeper layers, but also conversely, sediments and surface water can be polluted by contaminated ground water (e.g. with nearby hotspots on land). Hydrological models in combination with information on the chemical properties of the contaminants and the location-specific binding properties of the sediment and soil, as well as field measurements of fluxes or flows should be applied to evaluate those risks.

1.4. Surface water (re-)contamination

Contamination in sediments can pose a risk to ecology and humans *via* re-contamination of the surface water. This can be the result of advection or diffusion and of bioturbation or erosion of contaminated top-layers of the sediment. The relationship between sediment quality and this risk is complex, site specific, and requires a good understanding of the dynamics of the sediment/water system.

1.5. Eutrophication

Eutrophication can also be a strong driver for sediment remediation, but is normally not captured in the 'standard' risk assessment approach. In most cases, the risk assessment starts as a result of reports on the poor ecological quality of a water system, with an agreed need for diagnostic research (is sediment contributing to the problem?). An example is the Brazilian approach to distinguishing between nutrients and contaminants [24]. The next section will describe the current situation in European countries with respect to the use of risk assessment approaches for sediment management.

2. Approaches for assessment of *in situ* risks currently used in European countries

2.1. Ecological risk

Table 1 provides an overview of the ecological in situ sediment risk assessment approaches used in European countries. First of all, the management goals and the stage of implementation of these approaches differ greatly between the different countries. Only in a few European countries is sediment risk assessment performed from the perspective of a decision on whether or not to remediate (environmental dredging). However, in monitoring programs, basic physical and chemical methods are commonly used for sediment characterization. Physical investigations include particle-size distribution, organic carbon (OC) content, water content, etc. Physical analysis must be done by using the whole sediment, while chemical analysis is sometimes conducted on the fine-grained fraction, based on the reasoning that pollutants are enriched in this fraction. To link ecological effects/risks with sediment contamination, the bioavailable fraction of contaminants can be measured, but this is not common practice. The concentration of acid volatile sulphide (AVS) and simultaneously extractable metals (SEM) is also only occasionally measured to estimate the concentration of bioavailable heavy metals. In some countries, surveys of the benthic zoomacroinvertebrate community are carried out and/or bioassays are used as part of a more detailed risk assessment.

From Table 1 it is clear that the degree of implementation of *in situ* risk assessment approaches in legal frameworks in European countries is low. Between countries there is a large variation in how far biological endpoints are integrated in the assessment. The situation in a number of countries is given in section 2.5.

Table 1: *In situ* sediment risk assessment approaches used in European countries

Country	Management Goal	Sed Type[1]	Tiered?	Characteristics	Legal status: implemented in legislation?
Austria	-				
Belgium	Monitoring *in situ* sediment quality	F	No	Chemical + bioassays + benthic community	No
Czech Republic	Monitoring *in situ* sediment quality	F		Chemical	No
Denmark	-				
Finland	-				
France	Monitoring *in situ* sediment quality ecosystem restoration; flood management	F	Yes	Chemical, with biological endpoints as supplementary tools	Proposed
Germany	Monitoring *in situ* sediment quality	F		Chemical	No
Greece	Monitoring *in situ* sediment quality	H, M		Chemical	No
Hungary	Monitoring *in situ* sediment quality	F		Chemical	No
Ireland	-				
Italy	Monitoring *in situ* sediment quality	F		Chemical	No
Latvia	-				
Lithuania	-				
Luxembourg	-				
Malta	-				

Country	Management Goal	Sed Type[1]	Tiered?	Characteristics	Legal status: implemented in legislation?
Netherlands	Assessment of *in situ* sediment quality for prioritization of remediation sites	F	Yes	Chemical + bioassays + benthic community+ bioaccumulation measurements	Yes
Norway	Assessment of *in situ* sed. quality, also for prioritization of remediation sites	M, H (F)	Yes	Chemical + bioassays + biomarkers + bioaccumulation measurements	Yes
Poland	Monitoring *in situ* sediment quality	F		Chemical	No
Portugal	-				
Slowakia	Monitoring *in situ* sediment quality	F		Chemical	No
Slowenia	Monitoring *in situ* sediment quality	F		Chemical	No
Spain	Monitoring *in situ* sediment quality	F, M, H/E	Occasio nally	Chemical, bioassays (chronic and acute), bioaccumulation, biomarkers (under field and laboratory conditions, using caged animals and/or field collected organisms)	No
Sweden	Monitoring *in situ* sediment quality	M	No	Chemical, with biological endpoints as supplementary tools	No

Country	Management Goal	Sed Type[1]	Tiered?	Characteristics	Legal status: implemented in legislation?
United Kingdom	Monitoring sediment quality in relation to source control actions	F	No	Chemical	No
United Kingdom	Monitoring *in situ* sediment quality	M	No	Chemical + bioassays + benthic community + bioaccumulation measurements	No
United Kingdom	Assessing ecological quality and DM disposal options	F/M/H	Yes	Chemical + bioassay /benthic community + biomagnification potential	Under consideration

[1] Sediment type:
M = marine
F = freshwater
H = harbour/estuarine

2.2. Human health risk

Workgroup 5 of the SedNet project sought to determine the frameworks currently used in European countries for assessment of risks for human health. Most countries have no standardized risk assessment approaches that have been adopted in national legislation, e.g. for sediment management. In France, a framework for the assessment of human health risks has been proposed [25].

The Netherlands is the country which has gone the furthest in this work. Location-specific assessments of the *in situ* risks of sediment pollution in the Netherlands are mainly carried out in freshwater systems. For this evaluation, a tiered approach is followed (see also [26]), in which the first-tier assessment is a comparison of levels of priority contaminants with national intervention values for soil and sediment quality. These generic standards are based on the most critical information from either potential human toxicological or potential ecotoxicological risks, and do not discriminate between soil and sediments, type of land use, landscape ecology or site-specific conditions.

Actual risk assessment, as performed in The Netherlands, focuses on site-specific risks and considers current and intended or possible land use ([27], see also [28]). The risk assessment for human health requires the quantification of the actual exposure due to the use of contaminated soil and sediments. Total lifelong average exposure, expressed in mg/kg body weight per day, should not exceed the human toxicological Maximal Permissible Risk (MPR). The procedure to assess human risk is a tiered method described in detail by Otte *et al.* [29]. The methods can be regarded as conservative, because the underlying assumption is that actual risks are unacceptable, unless it can be proved otherwise. In addition, the choice for slightly conservative default model input parameters points to a careful approach.

The first tier concerns the calculation of total human exposure using the exposure model SEDISOIL (Figure 1). The model concept comprises the distribution of contaminants over the sediment, the surface water and the suspended particles using average partition coefficients. Formulae for the following exposure routes have been included in the model.

- Sediment ingestion
- Surface water ingestion
- Suspended particles ingestion (together with surface water)
- Fish consumption
- Dermal uptake via sediments
- Dermal uptake via surface water

The starting point for the model calculations is the total sediment concentration and the relevant sediment characteristics such as pH and clay and organic matter content. The first step is to select the exposure scenario, based on the actual site use. To facilitate model application, several 'standard' scenarios are included in the model: for recreation (e.g. swimming and sun bathing) and for fishing.

The calculated average lifetime exposure is compared with the human toxicological MPR (maximal permissible risk). When the exposure exceeds the MPR, one can decide on corrective measures or continue the investigation with additional measurements.

In the second tier, the assessment can be restricted to those contaminants that exceed the MPR levels in the first (conservative) tier. The essence of the second tier is the combination of site-specific field information and model calculations. Additional measurements are valuable when the calculated risk is related to exposure routes with an assumed high uncertainty. Uncertainty is caused by the model concept, input parameters and spatial and temporal variability. The exposure routes that were identified as crucial in the first tier guide the additional measurements in the second tier, which often focus on fish concentration, suspended particles and surface water.

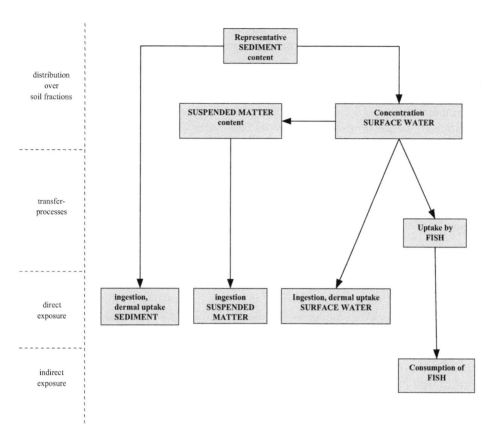

Figure 1: The SEDISOIL model for evaluation of risks of sediment contamination for human health

2.3. Risk for groundwater contamination

Many countries have a good deal of knowledge on groundwater dynamics on a local or regional scale in soils and sediments. Whether or not polluted sediments form a serious threat to the groundwater quality depends on many factors and makes a risk assessment rather complex. In the Netherlands, an assessment of in situ sediment quality also includes a quantitative assessment of the groundwater flow through the waterbed and the underground combined with a quantitative assessment of the (geo)chemical processes concerning the transport of micropolllutants from the sediment to the groundwater. Critical factors are the retardation of contaminants in the sediment and the underground. Another important element in the risk assessment is the modelling of advective and dispersive contaminant transport. Studies in the delta of the rivers Rhine and Meuse have provided new insight into these processes, and led to the development of a new numerical equation for dispersive transport, because in most models the magnitude of this was overestimated [28].

2.4. Risk for contamination of surface water

An important reason for concern about contaminated sediments is their possible negative influence on the quality of surface water. This risk is most relevant in old cases of sediment contamination and when the upstream sources of this contamination have been reduced. This could create the situation that contaminated sediments become the last (diffuse) source for (re)contamination of surface water. There are three mechanisms for recontamination of surface water by contaminated sediments:

• Diffusion
• Resuspension of sediment
• Erosion of sediment

The latter two are the most important mechanisms to be considered. For a quantification of these two pathways, thorough knowledge of the water system with regard to hydrodynamic (current speeds, waves) and morphologic conditions, combined with sediment contamination data are needed. The risk of surface water pollution depends on the morphological situation of the bed, with regard to the probability of erosion taking place. The risk assessment approach used in The Netherlands [28] uses as a benchmark that events (high discharge) resulting in an erosion of over 10 cm of the sediment layer are considered as possible risk situations when it concerns contaminated sediment present in the top layer. Alternatively, resuspension of contaminated sediments in shallow locations, e.g. by strong winds, may also have negative consequences for water quality. The next step in the assessment is an evaluation of the contribution of erosion or resuspension event to the pollution of suspended particles in the water column. If this exceeds 10%, then it is concluded that there is a risk for recontamination of surface water [28].

2.5. Some examples of integrated approaches currently used in Europe (ecological/human/groundwater/surface water)

Belgium (Flanders)
Sediment quality assessment in Flanders has been incorporated in a monitoring network by the Flemish Environment Agency since 2000. The focus is on freshwaters. Every year 150 locations are sampled, with specific locations resampled every four years, so in total 600 locations are included in the monitoring programme. The assessment is based on a triad approach [17,30]. Physical-chemical, biological and ecotoxicological assessment methodologies

are used, and an identical weight is assigned to each of the three assessments. The principle behind the classification of the watercourse sediments rests on an evaluation of the abnormality compared to a reference condition, so for each methodology a reference condition must be defined. This creates the possibility of classifying watercourse sediments in the absence of existing biological standards.

- **Physical-chemical assessment**
  The chemical parameters that are included in the assessment are nonpolar hydrocarbons (NPHCs), extractable organohalogens (EOX), the sum of the chlorinated pesticides (SOCP), the sum of seven PCBs (PCB7), the sum of six Borneff PAHs (PAH6), and heavy metals Cd, Cr, Cu, Ni, Pb, Hg, Zn and As. The concentrations are normalized to values for sediment with a standard granular composition and organic carbon content (see description of normalization in the section on The Netherlands). The site is classified based on the ratio to reference values. The sediments are ranked in classes based on the concentrations of the various contaminants. The sediment then receives an overall ranking based on the highest contaminant class ranking.

- **Ecotoxicological assessment**
  A battery of three tests is used for the ecotoxicological assessment. The battery consists of two pore water tests, namely a growth inhibition test with Raphidocelis subcapitata and an acute mortality test with *Thamnocephalus platyurus*, and one solid-phase test, namely an acute test with *Hyalella azteca*. The results are compared to results obtained with a reference sediment (with similar characteristics for grain size distribution, etc.). Based on the ratio, a classification is assigned. The ultimate ecotoxicological class is determined by the highest class of the two assessments (interstitial water and bulk sediment). The result is used as an estimate of the acute impact determined on aquatic life forms.

- **Biological assessment**
  Two indexes are used for the biological quality of watercourse sediments, namely a Biotic Sediment Index [31] and the percentage of mouth deformities of *Chironomus* spp. [32].
  Finally, the results are integrated based on the three classifications. This assessment method results in a rough indication of the sediment quality. To date, the results of this approach have not been used directly in sediment management. However, a method was proposed to use the information from risk assessment studies for the prioritization of remediation sites [33].

France

There is currently no framework applied in France which could be definitely attributed to this kind of risk assessment. *In situ* assessments are mainly related to (i) monitoring, (ii) ecosystem restoration or (iii) flood management.

(i) *Monitoring*: Until now, sediment quality monitoring for the protection of water bodies was done on the basis of analyses of priority pollutants and comparison to numerical SQGs. The use of toxicity bioassays is now seriously envisaged, following several demonstrative studies [34]. Two bioassays have been selected, i.e. *Chironomus riparius* (10 days, survival and growth) and *Hyalella azteca* (14 days, survival and growth) and will be applied after the formal adoption of a standard. Observations of invertebrate communities are also carried out at almost all the stations of the monitoring network, but their results are usually not matched with measurements of sediment contaminants. An attempt was made in 2001 to identify sensitive benthic organisms and reliable descriptive variables, and possibly underline contamination and biological response patterns [35]. This first study appears interesting, but should be extended with more powerful multivariate methods, and a more selective approach, as it appears that the impacts of contaminants are stronger in low current sections[1]. Ultimately, either the use of bioassays or matching benthic observations with sediment chemistry should help to consolidate or refine the existing SQGs.

(ii) Ecosystem restoration: there may be many reasons leading a local institution (municipality or group of municipalities) or a water manager to envisage a restoration of a degraded ecosystem, or some functionalities of that ecosystem. Dredging in this case appears as a technical solution among others, or as a part of the overall restoration process. In any case, the dredging project will be subject to authorization by the relevant authority or, depending of the volume, to a simple declaration. In the latter case, the project manager will have to describe all aspects of the project, while in the former he will have to provide a so-called impact study encompassing a broad range of issues.

(iii) Flood management: dredging may be proposed as a solution for managing floods in urbanized areas, or in the vicinity of dams. Again, the dredging project will be subject to authorization by the relevant authority or, depending of the volume, to a simple declaration.

The current guidance for these impact studies is rather open if not vague, and does not require the inclusion of ecotoxicological aspects. It is recommended that various management options, not only dredging, are considered [36]. The

---

[1] The standard method for invertebrate community assessment is based on sampling of various habitats; the above-mentioned approach looking for relationships between richness or abundance and sediment contamination did not discriminate between the habitats

relevant authority would generally ask for a focus on toxicological or ecological impacts if it knows beforehand, or suspects, that chemicals are present at the site of concern. A specific guidance for *in situ* sediment risk assessment is currently under development on behalf of the French Ministry of Transportation.

Germany

Sediment quality assessment to determine *in situ* risks in Germany is mainly based on chemical quality criteria. Three main systems are used: the assessment systems of the Joint Water Commission of the Federal States, LAWA (Länderarbeitsgemeinschaft Wasser); the Elbe River Water Quality Board, ARGE ELBE (Arbeitsgemeinschaft für die Reinhaltung der Elbe); and the index of geoaccumulation [37]. Whereas the LAWA and ARGE ELBE classifications are based on ecotoxicological effect levels of heavy metals, organic substances (industrial chemicals) and biocides, the index of geoaccumulation exclusively considers geochemical data, and does not take into account that heavy metals have different effects on organisms. Nevertheless it is still widely used.

The LAWA-System: The LAWA classification for sediment quality consists of four main and three sub-classes based on data for seven heavy metals, 28 organic chemicals, nutrients, salts and 11 sum parameters [38]. Quality class I reflects a natural or potentially natural environment with no xenobiotic substances measured in the sediment and with average geogenic (natural) background levels of heavy metals. Class II includes target values that are expected to guarantee high ecological protection. According to the Federal Environment Agency, it is not yet possible to designate target criteria for the protection of sediment-dwelling organisms, due to the lack of generally acknowledged methods. Therefore soil limit values in force under the Sewage Sludge Ordinance are adopted as water quality targets for the asset of Suspended solids and sediments [39]. These quality targets, which make up class II of the LAWA system, are considered to correspond to the good environmental quality category of the European Water Framework Directive. The limits between the higher classes (II–IV) are derived from a multiplication of the target values by a factor of 2.

The ARGE-ELBE-System: the classification according to the ARGE ELBE is structured similarly to the LAWA system but uses target values that have been decided upon by the International Commission for the Protection of the Elbe. It uses 27 priority substances and includes arsenic because of its special importance for the Elbe [38].

The Index of Geoaccumulation consists of a seven-tiered classification system, whereby the different classes derive from continued doubling of a background level. No biological considerations are involved.

Recently, recommendations were made for the use of an integrated stepwise approach combining toxicological, chemical and ecological information to assess and evaluate the quality of sediments [40]. A difference from approaches followed in most other countries is that bioassays are used as a trigger for further research steps, instead of the chemical data that are more commonly used. In Henschel *et al.* [41], a stepwise approach is described for an integrated assessment of ecosystem health effects and the consequences of sediment contamination for human health.

Italy

Presently, *in situ* sediment risk assessment approaches are not regulated in Italy and only recently some recent research and monitoring programmes have combined toxicological, chemical and ecological information. To assess the risks to biota, the Italian directive D.lgs. 152 [42] (to implement 91/271/CEE and 91/676/CEE directives) gives Water Quality Objectives, and sediment quality assessment is considered supplementary. In freshwaters, two procedures are provided:

- Basic analyses of water quality: anthropogenic impacts on biota are evaluated using the Extended Biotic Index (IBE), a biological species diversity index;
- Additional analyses: not mandated by law, but recommended to investigate short- or long-term effects in particular cases. Toxicity tests of concentrated water samples in *Daphnia magna*; mutagenicity and teratogenicity tests of concentrated water samples; algal development test; and tests of concentrated water samples in bioluminescent bacteria.

In addition, bioaccumulation analyses of priority pollutants (PCBs, DDTs and Cd) on muscle tissues of fish or on macrobenthos are recommended.

As for sediments, analysis of a number of metals and organic micropollutants are considered as supplementary and performed in particular cases to determine causes of environmental degradation of the water body. In case it is necessary to highlight short- or long-term toxic effects, the decree requires that bioassays are performed on sediment extracts, on whole sediment samples and on interstitial water.

To assess the state of the coastal/marine environment, bioaccumulation analyses of metals and organic pollutants (PAHs, PCBs, pesticides) in Mytilidae (*Mytilus galloprovincialis*) and Ostreoidea (*Ostrea edulis, Crassostrea gigas*) are of importance. If the indicator species are not present in the environment, Tellinoidea (*Donax trunculus*) and Veneroidea (*Tapes decussates, Tapes philippinarum*) should be used as alternatives. Additional analysis could be performed a) on key communities (phanerogame, reefs) to more completely characterize the ecological state of the environments, b) with bioassays to test short- and long-term effects of pollutants in different taxonomic groups (favouring autochthonous species for which standardized protocols exist). To classify the coastal/estuarine environment, there are no existing integrated approaches assigning ranks. The environmental classification will therefore be based on existing trophic indices of species diversity.

The Netherlands

In The Netherlands, sediment quality assessment has become part of routine monitoring programs, both in freshwaters and marine waters . Different sets of SQGs have been implemented for a chemical classification of sediment quality on a scale from 0 to 4 for freshwaters. These SQGs were to some extent based on ecotoxicological effect data [43,44]. More recently, SQGs have been developed for sediment quality assessment using species sensitivity distributions (SSDs) for specific chemicals [45]. Chemicals below their SQG value reflecting protection of 95% or more of the theoretically present species are considered to cause a tolerable degree of risk; chemicals higher than their SQG value reflecting protection of 50% or less of the species indicate potential high risk. The latter SQGs form, together with SQGs that have been derived specifically for assessment of potential human risk, the basis for the Dutch intervention value that plays a role in the soil/sediment protection act (see below).

Location-specific assessments of the *in situ* risks of sediment contamination in the Netherlands are carried out mainly in fresh water systems. The *in situ* sediment quality assessment is then part of a broader evaluation of the risks caused by sediment contamination, aimed at the question of whether the risks make sediment remediation necessary. For this evaluation, a tiered approach is followed:

I. First-tier assessment: comparison of levels of priority pollutants with national standards/guidelines. Chemicals measured routinely are mineral oil, chlorobenzenes, organochlorine pesticides, PCBs (standard group of seven congeners), PAHs (16 of EPA) and the heavy metals Cd, Cr, Cu, Ni, Pb, Hg, Zn and As. Contaminant levels are normalized according to the approach described by CUWVO [46], in order to compensate for differences in sorption characteristics between sediments[2]. Normalized contaminant levels are then compared with the Dutch sediment quality criteria (developed for first-tier assessment of risks for human health and ecosystems). According to the resulting classification, most contaminated sediments (exceeding intervention value(s): class 4 on a scale from 0 to 4) require a risk assessment (second tier).

---

[2] Standard sediment is defined as having a 25% particle fraction <2 μm and 10% organic matter on a dry weight basis

II. Second-tier assessment: If a priority contaminant exceeds the intervention value (indicating potential high risks, e.g. in the case of ecological risks: species-sensitivity distribution for the pollutant indicates that more than 50% of the theoretically present species might be affected), in tier 2 risk assessment is required in order to determine the need for specific actions (e.g., remediation). Not only contaminants exceeding the intervention value(s) are considered, but all contaminants present at elevated levels. Three main pathways are considered within this tier to achieve a complete risk assessment:

II.1. Human exposure: model calculations are carried out in order to quantify the extent to which humans (adults/children) can be exposed to contaminant via food consumption (especially fish consumption has to be considered) or via recreation activities in water. When the exposure exceeds maximum permissible risk criteria, actual risk is concluded. The model is based on general assumptions with regard to behaviour and diet of human populations. See also section 2.2 in this chapter.

II.2. Investigation of the risk of transport of contaminants from the sediment to groundwater or surface water. Model calculations are carried out in order to quantify the extent to which these processes occur. When contaminant fluxes (preferably calculated from field data) exceed high risk criteria, 'actual' risk is concluded. See also sections 2.3 and 2.4 in this chapter.

II.3 The third 'risk pathway' is about the (*in situ*) ecological risks of contaminants in the sediment. Traditionally, in the Netherlands risks for the ecosystem are evaluated using the triad assessment. In the Dutch version of the triad, bioaccumulation measurements are also considered, using the results of laboratory tests, or preferably by measurements using indigenous organisms [18]. Based on the *most sensitive* parameter, sediments are classified for the categories 'field observations' and 'bioassays' as either '−' (no effect/risk), '±' (moderate effect/risk) or '+' (strong effect/high risk). The goal is to elucidate the relationship between effects on macrozoobenthos and responses of bioassays which, in turn, can be related to levels of chemical pollution. For that purpose, chemical concentrations are converted into toxic units (TU): these are the ratio between the chemical's normalized concentration and the lowest NOEC[3] reported in the literature, among the bioassays included in the battery [18]. High risk is inferred when strong effects are observed in field surveys and/or bioassays that can be related to chemicals present in the sediment. More recently, it has been recommended that other lines of evidence such as chemical speciation, bioavailability, toxicity identification evaluation (TIE) procedures and field bioassays should be used to provide more evidence for cause–effect relationships.

---

[3] If sufficient data are available; the median value would be preferable

III. Integration of information and prioritization. When the data supplied from the second tier show that for none of the risk 'pathways' high risk is concluded in a site where one or more priority pollutant(s) exceed(s) the intervention value, there is no need for remediation.

Sites for which high risks are concluded (for at least one pathway), have to be prioritized for sediment remediation measures. The prioritization is based on cost-benefit analyses, taking into consideration the interests of stakeholders. Also different remedial options are compared for the expected risk reduction they result in. In the Netherlands, some experience exists with the use of multi-criteria analysis (MCA; also called Analytic Hierarchy Process – AHP; [47] for this purpose. MCA ranks sites based on the assessed risks. This method (described by [18]) is based on the same classification of results as described above. For *each* criterion (parameter), standard numerical values (scores) can be assigned to the effect/risk classes, from the value 1 for the class representing the strongest effect or highest risk, to for example 0.5 and 0.25 for the classes representing moderate risk and no risk (or remaining risk after remediation), respectively. Then the criteria are given a specific place and weight in a hierarchy. The scores are multiplied by the weight of the corresponding criterion and subsequently totaled bottom-up using a computer program, resulting in an overall score (in this example between 0.25 and 1). The approach can be used to give a numerical value for the overall risks for the ecosystem, but at a higher level of the hierarchy, information from human risk studies, ecological risk assessment, and estimates of contaminant mobility (transport) can be integrated. In the MCA, specific weights can be attributed to the different criteria (parameters) that are higher in the hierarchy, at branch points. This makes the method useful for decision makers, who must deal with all these aspects at the same time and therefore need integrated information.

Recent developments in The Netherlands with the assessment of (in situ) ecological risks

Recently the assessment of (*in situ*) ecological risks has been improved. Before a triad assessment is started, a risk estimation step is carried out using an ecotoxicological database (software model) and bioavailable contaminant concentrations. The resulting tiered approach is as follows:

 I. Comparison of priority contaminant levels (total contaminant levels in the sediment) with intervention values (indicating potential high risks, similar to tier I described above).

 II. In case intervention values are exceeded (tier I), the toxic pressure on aquatic organisms is calculated using the model OMEGA, and with bioavailable concentrations of contaminants in the sediment as the model input. The model will calculate the potentially affected fraction of species (PAF; [12]). The same methodology has been developed for the assessment of risks of soil contamination [48]. With this model, direct effects and effects as a result of food chain poisoning can be distinguished. In the Netherlands, mild extraction techniques with $CaCl_2$ or Tenax are used for measurement of the contaminant concentrations considered to be bioavailable [49] (see also [28]).

III. In case high toxic pressure is expected on the basis of bioavailable contaminant concentrations (tier II), the third step consists of a Triad sediment quality assessment. The selection of parameters and classification criteria for a new Triad approach is ongoing and aims to relate better with water framework directive quality objectives (trying to link sediment and water quality).

The new ecological risk assessment approach can be integrated with information from assessments of human health risks and/or risks from transport of contaminants to surface water or groundwater in a last tier, as described above. There remains one major limitation in the new ecological risk assessment approach: only a limited group of chemicals is considered in the first two tiers. To this point, this was justifiable, because the sediment quality assessment has been part of the legal framework of the soil protection act that focused on a group of known contaminants (known historical river pollution). However, it is clear that ecological problems can be caused by the (combined) toxicity of many more contaminants than those that are routinely measured. In the near future, sediment ecological risk assessment will become a tool supporting the implementation of the water framework directive (WFD). Sediment quality assessment may be necessary when the chemical and/or ecological objectives for a water body are not met, leading to the question "is the sediment quality the limiting factor for reaching ecological objectives?" In this case, of course all chemicals present have to be evaluated for the possible risk they may cause. Therefore, the ideal (stepwise) approach to answer this question would be to start with sensitive wide spectrum biotests (see also section 4).

Norway
Fresh and marine waters. The Norwegian State Pollution Control Authority (SFT) in cooperation with the Norwegian Institute for Water Research (NIVA) and the Norwegian Geotechnical Institute (NGI) have developed a national framework for conducting *in situ* risk assessments of contaminated sediments [50,51]. The main objective of the framework is to assist in the prioritisation of harbour and fjord areas for remediation. However, the framework can also be modified and used for freshwater sediments.

The risk assessment framework is built up in three tiers:

- Tier 1 constitutes the comparison of measured data with sediment quality criteria (SQC) for acceptable risk. The SQCs are established for concentrations of priority pollutants in sediments as well as for the toxicity of sediments. Toxicity is included in order to ensure that any potential contributions of toxic substances that are not identified in a chemical characterisation are also taken into consideration. SQCs are established for the following pollutants: mineral oil, TBT, PCBs (sum of 7 congeners), PAHs (10 of the 16EPA) and the heavy metals Cd, Cr, Cu, Ni, Pb, Hg, Zn and As. Toxicity is investigated using the marine algae Skeletonema costatum (pore water and organic extract) as well as the DR-Calux assay or EROD activity.

- Tier 2 comprises three independent assessments with a focus on site-specific conditions; i) risk for contaminant transport, ii) risk to human health and iii) risk to the ecosystem. Contaminant transport is assessed by calculating contaminant fluxes based on field data. The human exposure assessment is based on the Dutch SEDISOIL model, modified for Norwegian conditions. Risk to the ecosystem is assessed using the toxicity test data from Tier 1 as well as an evaluation of sediment toxicity (using crustacean *Corophium* or polychaete *Arenicola marina*) and a bioaccumulation test (using polychaete *Hediste diversicolor* and gastropod *Hinia reticulata*).

- Tier 3 is an extensive assessment with a focus on collecting field measurement data to verify and clarify the calculations conducted in Tier 2. The contents of Tier 3 are customized on a case by case basis.

The United Kingdom
There has been considerable research and development in the field of sediment risk assessment in the UK, but not much uptake in a regulatory sense, particularly for freshwater sediments. The UK had a very active period of sediment research in the mid-1990s, resulting in broad reviews of approaches to risk assessment (e.g., [52,53] as well as establishment of the National Marine Monitoring Programme [54]).

In the UK, *in situ* freshwater sediments are only routinely assessed for environmental quality within the framework of the EC Dangerous Substances Directive (76/464/EEC), together with the Water Resources Act 1991, both of which require control over inputs of dangerous substances into water. Specifically for sediments, List 1 Dangerous Substances are monitored in sediment or biota at sites proximate to dischargers that discharge those substances, under a standstill provision, which states that it is necessary to demonstrate that the levels of a particular substance, present in either sediments and/or biota, do not increase significantly with time [55]. In addition, *ad hoc* investigations of freshwater sediment quality are conducted, related to site-specific issues such as contaminated land, navigable waterways management, water quality problems and academic interest. Aside from draft sediment quality standards for dioxins and furans in England and Wales [56], there are no freshwater standards for sediment assessment at this point in any part of the UK. In the past few years, the Environment Agency has been reviewing their policies related to sediment assessment and management. They commissioned two reviews to advise them. The Environment Agency reviewed the derivation of sediment quality guidelines [57] and the nature and extent of sediment issues, including in situ sediment risk assessment [58]. In a more recent report, Apitz *et al.* reviewed the ways in which SQVs have been implemented within sediment assessment and management frameworks in other countries so that an implementation strategy appropriate to England and Wales can be developed for the Environment Agency [59].

The authors of a recent document by the Environment Agency stress that SQGs are insufficiently reliable to support automatic regulatory action should guideline concentrations in sediments be exceeded [57]. Exceedance of SQGs should always trigger investigatory actions that seek to confirm or deny the predicted risk. It was argued that SQGs could be useful in the UK, provided (i) they minimize false negatives (type II error) and (ii) their exceedance must not be the sole reason for regulatory action [57]. They should rather be used as a first screening, along with considerations of background concentrations, which is consistent with the conclusions drawn in another document by the Environment Agency [58].

Although the Environment Agency supports tiered approaches to sediment assessment, they do not propose a definite framework at the moment, but recommend developing and validating an approach for the UK [57,58]. It was recommended later by Apitz *et al.* that the Agency should begin to roll out use of SQVs within separate, tiered risk assessment frameworks for assessing ecological quality and dredging activities [59]. They further recommended that other lines of evidence such as chemical speciation, bioavailability and bioassay studies, plus ecological surveys, should also be used in these frameworks if they can add value to management decisions.

In marine and estuarine waters, the UK's National Marine Monitoring Programme (NMMP) is a well-developed programme that monitors sediment quality, essentially using a triad approach. The Marine Pollution Monitoring Management Group (MPMMG) is a management group for the programme, with representation from all government organizations with statutory marine environmental protection monitoring obligations.

The NMMP was developed in response to OSPAR as well as several EC Directives The NMMP Phase 2 focuses on stable depositional sediment sites (approximately 110 sites) and evaluates: sediment chemistry, benthic communities, bioaccumulation, and ecological effects methods. It is also anticipated that NMMP data will be used to fulfill some of the monitoring requirements of the Water Framework Directive[4]. Initially, the main objective of the programme was to describe marine quality around the UK through spatial surveys (phase 1), but it has now shifted to detecting with appropriate accuracy long-term trends in physical, biological and chemical variables at selected estuarine and coastal sites (phase 2). Other objectives include support for consistent standards in national and international monitoring programmes for marine environmental quality (for example, EC Directives, OSPAR) and making recommendations on how new analyses and techniques are best implemented in the United Kingdom. Overall, the aim is to produce reports providing overviews of the spatial (NMP holistic report 1998) and temporal distributions (every 3 years from 2002) of these variables and their inter-relationships.

Spain

Freshwaters. No sediment quality assessment other than the use of chemical measurements is recommended for the management of sediment polluted in fresh water from Spain. However, no SQGs are recommended to evaluate chemical measurements in sediments. Most of the assessments carried out in these areas are based on the geochemistry of sediments. Some research includes the use of biological assays such as the commercial Microtox,® but only as complementary information. Most of the studies that have used both chemical and biological data were forced by the Spanish legislation related to hazardous material, but were not part of an integrated approach for sediment quality monitoring in these environments. The biological assays used for the assessment of hazardous material (mammalian tests, no freshwater organisms) are inappropriate for determining the quality of sediments.

Estuarine and marine waters. Most of the regulatory agencies both from the Central and Autonomous Governments follow a typical triad schema to determine sediment quality in these areas. It is important to note that it is not a weight of evidence approach (e.g. the triad as described by Chapman [60]), but a general monitoring using the main idea from the triad: quantification of pollution based on the assessments of the contamination (enrichment of anthropogenic substances) and the effects associated with those polluted sediments. It is not mandated by any law but has been adapted by most regulatory agencies. The triad approach has been used in Spain since about

---

[4] Full details of all sites together with methodologies, sampling schedules and frequencies are provided in the NMMP2 monitoring manual – 'The Green Book' at www.marlab.ac.uk/nmpr/nmp.htm.

1992–1993 [61]. From the results obtained, different sediment quality values (guidelines) were derived by comparing them to those obtained in a similar triad application carried out in San Francisco Bay (USA) and reported by DelValls and Chapman [61]. These sediment quality guidelines haven been used following a tiered approach for the monitoring of the impact provoked by a mining spill (Aznalcóllar, April 1998) in some coastal areas located in the Gulf of Cádiz [62,63].

The SQGs reported by DelValls and Chapman [61] were used to evaluate the concentrations of certain heavy metals from the mining spill in different areas of an impacted estuarine and coastal environment [64]. Briefly, an ecological risk factor was derived by calculating the ratio of the measured concentration of the metals to their respective SQGs (Tier I). Those areas with ratios higher than 1 in some metals from the mining spill were considered as potentially impacted (polluted) and then tested using a battery of toxicity tests (both sediments and waters) and compared to other clean stations in the area with ratios clearly below 1 (Tier II). Based on the results of Tier II, stations were selected for application of Tier III, the sediment quality triad. The triad results were compared to those of an area known for its contamination by mining activities (e.g. Huelva). Such use of SQGs and integrated *in situ* sediment quality assessment approaches is under consideration by different regulatory agencies and is being discussed by expert commissions for the monitoring of the impact in the Galician coasts after the oil spill provoked by the tanker *Prestige*.

3. Risk assessment for dredged material management in European countries

3.1. Ecological risks

Over the past decades, different risk assessment approaches have been implemented in dredged material (DM) management in Europe. These approaches are listed in Table 2. As for the *in situ* sediment risk assessment, basic physical (particle-size distribution, TOC, water content, etc.) and chemical methods are commonly used for sediment characterization. Physical analysis is on whole sediment, while chemical analysis can be done on the fine-grained fraction. The use of measurements of the bioavailable fraction of contaminants in the sediment is only relevant when the future disposal or re-use conditions can be mimicked. Bioassays are often included in the sediment quality assessment or added as a second tier. However, pass or fail criteria are often still being debated. Part of the risk assessment for dredged material management could be an assessment of the effects of the dredging operations and/or of the proposed disposal/re-use options.

Sediment management strategies fall into five broad categories that are selected based on an evaluation of site-specific risks and goals: (1) no further action (NFA), which is only appropriately applied if it is determined that sediments pose no risk; (2) monitored natural recovery (MNR), based on the assumption that while sediments pose some risk, it is low enough that natural processes can reduce risk over time in a reasonably safe manner; (3) in situ containment (ISC) in which sediment contaminants are in some manner isolated from target organisms, though the sediments are left in place; (4) in situ treatment (IST); and (5) removal via dredging or excavation (followed by ex situ treatment, disposal, and/or re-use). Figure 2 illustrates various sediment containment and disposal scenarios from in situ to ex situ examples. The information required to evaluate or compare each of these options is different, and any assessment should be designed to evaluate and support the specific management goals and potential remedial options. Careful planning is necessary to ensure that sampling and analysis plans are designed so they can address these disparate needs in a meaningful and comparable way [65].

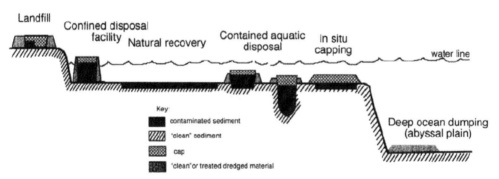

Figure 2: Containment and disposal options for dredged material (source: [66])

Table 2: Assessment of dredged material quality and impact. "Management goal" details the disposal option for which assessment is applied; Abbreviations in "Sed Type" comprise M: marine; F: freshwater; B: brackish; H: harbour/estuarine; "Chemical Assessment" refers to general/signalling possible ecological risks; BA: Bioassays; BAM: Bioaccumulation measurement; BCA – Benthic Community Analysis. Legal status refers to implementation into legislation.

Country	Manage-ment goal	Sed. Type	Tiers?	Chemical Assessment	Sediment toxicity testing	BAM	BCA	Legal status?
Austria	-							
Belgium	-							
Denmark	Disposal at sea			Contaminant concentrations are compared with those in sediment of relocation site	No	No	No	Only chemical para-meters imple-mented
Estonia	-							
Finland	-							
France	Disposal on land	F	Yes	In 1st tier, calculating quotients with SQGs consen-sus-based (mostly SQG from literature based on empirical data)	2nd tier with (chronic) sediment toxicity tests; 3rd tier with (chronic) toxicity tests using terrestrial organisms	As 3rd or 4th tier	No	No
France	Disposal in open pits	F	Yes	In a 1st tier, calculating quotients with SQGs (SQGs are consensus-based and mostly from literature based on empirical data)	2nd tier with (chronic) sediment toxicity tests; 3rd tier with (chronic) toxicity tests using sediment and aquatic organisms	As 3rd or 4th tier	No, because approach is inten-ed as prog-nostic	No

Country	Management goal	Sed. Type	Tiers?	Chemical Assessment	Sediment toxicity testing	BAM	BCA	Legal status?
Germany	Relocation / Open water disposal	F/H	No	Contaminant concentrations are compared with those in sediment of relocation site	Acute toxicity tests; toxicity evaluated by pT index	Optional in medium risk cases	Yes	Yes (ministerial decrees)
Greece	-							
Hungary	-							
Ireland	-							
Italy	For relocation in Venice Lagoon	B		SQGs based on background levels; national use: SQGs based on empirical data are under development	Under development	No	No	Yes
Latvia	-							
Lithuania	-							
Luxembourg	-							
Malta	-							
Netherlands	Free disposal in marine waters	H/M	No	SQGs partly based on EP assumptions and toxicty data	BAs in effect monitoring	(Occasional)	(Occasional)	Yes
Netherlands	Disposal by relocation in floodplains (on land/in pits)	F	Yes (tailor made)	As a first tier, comparing concentrations with SQGs	Optional	Optional	Optional	Yes

Country	Manage-ment goal	Sed. Type	Tiers?	Chemical Assessment	Sediment toxicity testing	BAM	BCA	Legal status?
Norway	Confined and free disposal on case by case basis	H/M	Yes (tailor made)	SQGs and flux calculations specific to the disposal option	Occasionally	Occa-sionally	Occasion ally	Yes
Poland	-							
Portugal	-							
Slovakia	-							
Slovenia	-							
Spain	Free disposal and/or beneficial uses	H/M	Yes	Comparing to SQGs and background levels	Three bioassays (chronic and acute)	Under develop ment	No, but specific guide-lines deve-loped for ports in Basque country	No, but guide-lines exist
Sweden	-							
Czech Republic	-							
United Kingdom	Free disposal in marine waters	H/M	No	Comparing concentrations with SQGs (but no scientific background for SQGs)				Yes

3.2. Risk for contamination of groundwater or surface water

Considering the diversity in management option for disposal of dredged material, also different (tailor-made) risk assessment approaches exist for the evaluation of groundwater or surface water contamination. In a number of countries, the beneficial re-use of (treated) sediments depends on specific requirements with regard to leaching characteristics. These techniques can also be applied to (untreated) sediments for which the preferred option is to relocate them in the environment. Also, modelling of contaminant behaviour during and after relocation can be part of the risk assessment.

3.3. Some examples of integrated approaches currently used in Europe

France
The political framework for dredged materials is still under discussion for freshwaters. From a legal point of view, these materials are classified as wastes, but not necessarily as hazardous. Another confusing issue is the destination of the materials – a deposit on soils is subject to a different set of regulations than those applying to a deposit in waters. There is thus a need for guidance at various levels of the management process; some aspects of guidance, e.g. for the overall management process, have been introduced [36] and should now be completed by more specific frameworks for the evaluation of the dredged materials. On behalf of the Ministry of Equipment and Transportation, such a framework was recently proposed for the ecological side of the assessment [67].

Disposal on soil
If the deposit is located close to a river or a canal, contaminant transfers to the surface water may occur, or to the surrounding soils, and to the groundwater. Organisms of concern include plants and aquatic species. The following assessment endpoints have thus been proposed:
• The deposit should not disrupt the germination or growth of plants, in particular those of agricultural value;
• Run-off waters should not affect aquatic species;
• Finally, it should not degrade the groundwater quality, i.e. for drinking water purposes.
In this scenario, stressors are represented by two types of water samples: excess water (mixture of overlying and pore water) collected on the deposit, which will support the transfers to the surrounding soils or the surface waters, and water obtained from elutriate assays in unsaturated packed columns. The tested assumptions are assessed with bioassays on bacteria (Metplate®), unicellular algae, a pelagic crustacean (*Ceriodaphnia dubia*), amphibians, and vegetables (lettuce, maize, etc.). The soil macrofauna and microflora have not yet been considered, but should be in the future versions of this protocol.

Disposal in water
In case a disposal site is constructed as a cross-section of the alluvial groundwater, the water will flow through the dredged material deposit and

contaminants may be eluted over time. Aquatic species can also be affected at the time of deposition, by direct exposure to pollutants dissolved in sediment pore water. Benthic species may be affected in various ways, in particular when they colonize the deposit. The following criteria for allowing disposal in water have thus been proposed:

- The deposit should have no effect on the structure and abundance of benthic invertebrates in the location;
- It should have no long-term effect on pelagic species;
- It should not cause groundwater pollution, as such disposal sites are in fact cross-sections of shallow alluvial groundwater.

A fourth assessment endpoint should be introduced, regarding health risks for recreational uses, including fishing, but this endpoint was not implemented in the current version of the approach. The analysis phase includes aquatic bioassays (bacteria-Metplate^{TM}, algae, microcrustaceans *Ceriodaphnia dubia*, rotifers *Brachionus calyciflorus*), and leaching assays in columns under ascendant flow.

Germany

For the Federal waterways in Germany, comprising over 90 % of all navigable waters, the Federal Ministry of Transport (Bundesministerium für Verkehr BMV) and its subordinate authorities are responsible. Conceptual guidance and project monitoring with regard to environmental aspects are covered by the Federal Institute of Hydrology (Bundesanstalt für Gewässerkunde BfG). All other inland waterways are under the responsibility of the Länder (Federal states), which have their own guidelines and recommendations (see also: [68])

Two directives apply to the federal waterways: the Directive for the Management of Dredged Material in Federal Inland Waterways for freshwater areas [69], and the Directive for the Management of Dredged Material in Federal Coastal Waterways for coastal waters up to the freshwater limit [70].

Federal waterways in the freshwater zone. The evaluation of dredged material is based on physical (e.g. grain size, water and organic content), chemical (seven heavy metals and arsenic, 28 organic compounds, four organic tin-compounds), biochemical (oxygen consumption, nutrients) and ecotoxicological (test with green algae, luminescent bacteria, dapnia magna) characterization.

Criteria for the relocation of dredged material include a comparison of the characteristics of the sediment with the characteristics of the proposed deployment area, where chemical and biological parameters are generally determined, with biochemical and ecotoxicological measurements being carried out only as an exception.

The decision-making system for the fate of the dredged material foresees three possible outcomes based on chemical criteria:

1. No measured contaminant shows a concentration higher than 1.5 times the average concentration at the relocation site: relocation is possible.
2. No measured contaminant shows a concentration higher than 3 times the average concentration at the relocation site: relocation is possible, if no alternative measures can be applied and if no adverse effects are expected.
3. At least one measured concentration exceeds the average concentration of that substance at the relocation site by a factor of 3: relocation is not possible. Measures other than relocation must be applied.

In case ecotoxicological criteria are applied, bioassays are performed in order to assess the toxicity of the dredged material [71,72]. The toxicity class is evaluated according to the pT-value of the most sensitive organism (pT = negative logarithm of the dilution factor necessary to reduce the effect below threshold). So far, for biochemical parameters no action limits are available.

Federal waterways in the coastal zone. The concept for the coastal zone area, based on international guidelines, is in principle similar. The main modifications are a stronger emphasis on nutrient concentrations and the use of two interim action levels for phosphorus, nitrogen and the chemical substances that are to be measured.

Current discussion on revisions of these risk assessment schemes focus on the following aspects:

- Basing decisions on the outcome of only one (the most sensitive) out of three ecotoxicological tests results in high uncertainty of the final conclusion on one side and in a low comparability of data for different sediments on the other because results from different test systems may be used for making the final decision.
- Even though relocation is the favoured action with regard to dredged material, assessment of sediment bound toxicity by direct contact tests is not included in the decision framework. A high toxicity resulting from particle-bound contaminants may not be detected in the elutriates but may still pose a danger during relocation.

Local assessment schemes. The port authority of Hamburg carries out sediment tests prior to dredging. Although ecotoxicological measurements have accompanied the sediment testing for 10 years now, decisions about the fate of dredging material are made almost solely according to chemical data.

Italy

For the assessment of sediment quality and management of disposal of coastal/marine dredged material, ICRAM is defining National Sediment Quality

Guidelines. The use of bioassays and bioaccumulation tests is planned in the process of quantifying the environmental risk. Pilot projects on dredged sediment treatments, focused on sustainable re-use in the environment, are ongoing.

For some specific local areas, quality criteria for sediment management exist. In the Venice Lagoon for example, a set of local quality criteria exists for evaluating the concentrations of metals, PCBs, PAHs, HC and pesticides [73]. Basically there are three classes of sediments that can be used inside the lagoon with increasing cautions related to the levels of pollutants. Sediments of Class A (good quality) can be reused for morphological restoration of the lagoon. Sediments of Class B (medium quality; easily manageable) can be reused in the lagoon islands but need to be permanently confined to prevent the release of pollutants into the water. Class C (poor quality; careful handling required) can be re-used only in parts of islands that are permanently dry (no risk of flooding) and are not subject to erosion. Above the level of Class C, it is necessary to send the dredged sediments to a landfill on the mainland.

No bioassays are required for the evaluation of the quality of dredged sediments. However, since the total pollutant concentrations do not give information on the availability and toxicity of contaminants, much emphasis has recently been put on the ecotoxicological evaluation of sediment quality. Therefore the use of bioassays and the comparison of the levels of contaminants in surficial sediments with reference values (ERL: Effects Range-Low, and ERM: Effects Range-Medium) such as those proposed by US-NOAA (National Oceanic and Atmospheric Administration), is encouraged.

The Netherlands
Freshwaters. Heavily contaminated sediments must be transported to confined disposal sites or should be treated in order to make beneficial reuse possible. The management of slightly or moderately contaminated dredged material from freshwaters (Class 1-3) in the Netherlands is still being debated. Until now, class 1 or 2 dredged sediments have mostly been put on land or have been used in construction works (subject to certain criteria in terms of chemical composition). A risk assessment framework is being developed that predicts the risks of prolonged free disposal of dredged sediment, considering the specific function of the land where the sediment is to be disposed of.

Marine waters. For the assessment of marine dredged material disposal in the coastal waters of The Netherlands, a few years ago a new sediment quality assessment approach was proposed, the Chemistry-Toxicity Test (CTT; [74,75]). Three bioassays were selected for routine application in the CTT approach, viz a mud shrimp toxicity test (*Corophium volutator*), a bacterial test (Microtox Solid Phase) and the DR-CALUX assay, which reacts specifically to dioxin-type compounds. The basic idea of the CTT approach was that, in order to allow sediment relocation in the Dutch coastal waters, sediment quality guidelines need to be met both for the concentrations of a list of chemicals and for the degree of effect observed in the bioassays. The attempts to implement bioassays in a regulatory framework (pass/fail criterion) for relocation of dredged material in coastal waters raised much discussion with respect to the robustness of the tests and economical consequences for maintenance dredging in Dutch harbours. This recently led to the recommendation not to use bioassays as pass/fail criterion in the CTT approach, but instead to give bioassays a role in monitoring programs at the relocation site [76]. Information from monitoring studies, in particular from the assessment of dredged material quality will then be used as a first step in the identification and source control of effect-causing contaminants.

Norway
There is currently no Norwegian national framework for conducting a biological effects-based sediment quality assessment for the *ex situ* risks of dredged sediments. Norway is in a unique situation where the need to carry out dredging activities is limited. Deep fjord systems and the absence of large rivers limit the accumulation of sediment in most harbour areas. When dredging or dumping activities do need to be carried out, a permit must first be obtained. Dredging and dumping activities are specifically administered by the regional commissioner according to the regulation which controls dredging and dumping operations in the sea and watercourses [77]. The permit application must include 'all information that is necessary to assess permit approval. . ., including characterization of the material and site conditions. . .' However, the sediment characterization requirements vary from region to region. Normally the sediment grain size distribution and chemical analysis of heavy metals, PCB, PAH and TBT are used to evaluate the application for dredging or dumping. When assessing the analytical results, it is common practice to use the upper limit of sediment quality class II to determine whether free aquatic disposal is acceptable.

United Kingdom
Disposal at sea. In England and Wales, sea disposal is regulated nationally by the Department of Environment, Food and Rural Affairs (Defra), but many of the decisions are driven by policy decisions made within OSPAR. Defra controls these activities relating to sea disposal through a system of licences under the Food and Environment Protection Act (FEPA) 1985. This Act provides a licensing system for the deposit of substances and articles from

vehicles and vessels, etc. in tidal waters below the level of mean high-water springs.

Sea disposal licences are only issued after detailed scientific assessment [with the support of the Centre for Environment, Fisheries and Aquaculture Science (CEFAS) who advise Defra] of the potential environmental impact, with particular regard to the need to safeguard marine conservation sites, fisheries and other uses of the sea. Prior to this year, the assessment procedure focused on (1) review of sediment data (physical quality and chemical quality relative to action levels) from the area proposed for dredging and (2) information about the sea disposal site and its ability to assimilate the materials proposed for disposal. As of this year, bioassay data are being collected in parallel to sediment chemistry data (i.e. *in situ* sediment quality assessment). In addition, CEFAS are testing a new dredged material disposal assessment decision tree (Murray, pers. comm.), which is both rule- and risk-based, providing a tiered assessment procedure that considers not only environmental risks but also beneficial uses for dredged materials proposed for disposal.

In summary, to assess the potential effects of contaminants, firstly the physical properties of the sediment are assessed. Secondly, the sediment chemistry of materials proposed for disposal at sea are assessed using action levels (applied by CEFAS) to give an indication of the potential for impacts. A standard suite of chemicals is used in the first instance and augmented as needed for site-specific conditions. CEFAS has an assessment procedure that involves two action levels (Action Level 1 and Action Level 2). Below Action Level 1 the material is usually suitable chemically for beneficial use or for sea disposal, while below and above Action Level 2, further assessment will be required before a licence for either sea disposal or beneficial use is issued. Action level figures are not pass or fail criteria, however, as the approach used by CEFAS is one of weight of evidence. Using the physical, chemical and bioassay data in parallel to make decisions about the suitability of dredged materials for sea disposal will permit CEFAS to collect enough data to evaluate this new approach, and the decision tree will be modified in light of CEFAS's findings. A recent report for the Environment Agency for England and Wales reviewed the ways in which sediment quality guidelines have been implemented within sediment assessment and management frameworks (including DM disposal options) in other countries so that an implementation strategy appropriate to England and Wales can be developed for the Environment Agency [59]. It was recommended that the Agency should begin to roll out use of SQVs within separate, tiered risk-assessment frameworks for assessing ecological quality and dredging activities (see also section 2.5).

Disposal on land (spreading and in landfills). Maintenance dredging in inland waterways is subject to limited environmental legislation, as reviewed in Bates and Hooper [78]. Capital dredging is subject to the same controls as maintenance dredging, but in some cases requires a full environmental assessment (which has the scope to include ecological risk assessment, but does not usually do so).

Under the Waste Management Licensing Regulations (WMLR) 1994, all disposal of dredged material not qualifying for an exemption must be licensed. Management of sediment through spreading on land under exemptions is regulated by DEFRA, through the WMLR. Generally, only site history and sediment chemistry data are used, but this is under review. For dredged materials that are very heavily contaminated, the Special Waste Regulations (1996) might come into play, and these are specify chemically driven assessment procedures.

Spain
Freshwaters. Only limited data on chemical concentrations in dredged sediments from freshwaters in Spain are used for management decisions. The importance of chemical data is ignored, partly because the quality of freshwater sediments to be dredged is assumed to be comparable to that of the soil/sediments along the shores of the rivers. Moreover, most dredging operations lead to the disposal of the dredged material on these shores.

Estuarine and marine waters. SQGs have been recommended for priority contaminants that should provide the criteria to chose disposal options for estuarine and marine dredged material in the coastal waters of Spain. Category I (free disposal to sea, only mechanical effects) will be established for dredged material with contaminant concentrations equal to or lower than the action level 1. If the concentrations are higher than action level 1 but lower than action level 2, the dredged material is considered category II (disposal to sea under controlled conditions followed by an integrated environmental risk assessment as described for the *in situ* sediment quality assessment used in Spain). Dredged material with concentrations higher than action level 2 are considered category III (free disposal not authorized; confined disposal required). In case of remediation (treatment) of the material it could be re-assessed after treatment and then disposed under the recommendations of category II.

Recently, research has been carried out with the aim of incorporating the use of sediment toxicity tests under a tiered approach for the characterization of dredged material in Spanish ports. The species used as test organism in the test battery included different benthic and pelagic taxa reflecting different exposure routes, including interstitial water, elutriate and whole sediment. Some screening (commercial) tests were included as well. In the near future, new chemical guidelines and the use of sediment toxicity tests are being discussed and should officially become part of the dredged material management in harbour and marine port areas in Spain.

4. Towards a standardized approach

When the European Water Framework Directive (WFD) came into force, it introduced the management of water on a river basin scale. Respecting the fact that the elements in the system are connected and that efforts to maintain and improve the ecological status of water bodies need to be coordinated, risk management should also be accomplished and priorities set on that scale. It makes no sense if a downstream manager is extremely cautious while the upstream manager is very pragmatic and sets other priorities, or the other way around. Management constraints in the river basin and the receiving coastal zone should be focused on actions that are most effective on the scale of the river basin, including the coastal zone. This should be facilitated amongst others by harmonized risk assessment approaches. It is clear that these assessment schemes must be management goal-orientated. Sediment management is a key issue in the maintenance of rivers and waters in many European countries. A significant problem in sediment and dredged material management is sediment contamination. Contamination in sediments can pose a risk to ecology and humans and may disperse to surface water and groundwater. However, the relationship between sediment quality and these risks is complex and site-specific. Sediments can stay in place when risks of contaminants present are acceptable. When dredging is required for other reasons than environmental risk, such as for maintenance or construction works, the dredged material has to be removed from a specific site and quality/risk assessment is necessary to guide the reintroduction of the material elsewhere (confined disposal, beneficial use or relocation in the water system itself).

At this moment, there is no agreement between the responsible authorities on a general scheme for risk assessment of contaminated sediments. One of the major reasons is that the legislation on contaminated sediments (*in situ*, or after dredging, *ex situ*) differs greatly between the European countries. In general, data are available for only a minor percentage of the chemicals that are released into the environment from point or diffuse sources. It is obvious that chemical

SEDIMENT			DREDGED MATERIAL	
(In situ risks)			*(Ex situ* risks)	
Tier 1	Tier 2	Tier 3	Tier 1	Tier 2
Screening BAs	BAs with endpoints of chronic toxicity	BAs with more representative species In situ BAs	Screening BAs	BAs with endpoints of chronic toxicity TIE methods
Ecological indicators of stress	Biological observations: BCS, BCA Bioaccumulation (analysis in biota / effect monitoring)	Advanced effect studies (health indicators; reproductive success) Demonstration of relation: sed. quality - ecological WFD objectives	Ecotoxicological models: predictions of effects	Monitoring program at relocation site
Physical parameters and chemical analyses (bioavailable concentration)	Bioavailability	Additional analyses: advanced list of chemicals TIE methods	Physical parameters and chemical analyses (bioavailable concentration)	Product quality requirements (evaluation of re-use options) Waste classification[5]
Selection of risk assessment tools for other risk pathways (if relevant)	Human health (risk model)	Analysis of contaminants in food (e.g. fish) Pathogen tests, skin tests	Selection of risk assessment tools for other risk pathways (if relevant)	Human health (risk model)[6] Pathogen tests, skin tests
	Groundwater (risk model)	Measurements in field		GW protection: leaching tests, modelling
	Surface water (risk model)	Measurements in field		Surface water contamination: leaching tests, modelling

Figure 3: Proposed general scheme for tiered assessment of *in situ* and *ex situ* risks of contaminated sediment and dredged material (BA – Bioassays; TIE – Toxicity Identification Evaluation, BCS – Benthic community survey, BCA – Benthic community analysis: GW – Groundwater)

---

[5] e.g. leaching tests according to 2003/33/EEC regulations
[6] e.g. evaluation of accumulation via food consumption

substances in the environment have ecological importance due to their effects and not simply from their presence. Thus it can be concluded that the investigation of biological function and the condition of the ecosystem is needed in addition to chemical analysis to prevent undesired burdens for the aquatic environment and to achieve sustainable management. Depending on the situation, information on risks to human health and risks caused by the transport of contaminants to ground water or back to surface water can also be necessary for proper risk management.

One possible approach to integrating different investigation tools into a general tiered scheme for risk assessment (of both *in situ* and *ex situ* risks) is presented in Figure 3. In the sections 4.1 and 4.2 (below) *in situ* risk assessment and risk assessment for dredged material management will be further discussed on the basis of Figure 3.

4.1. In situ sediment risk assessment

Ecological risk

For the assessment of the *in situ* ecological risks, some recommendations can be made for the development of a general, tiered risk assessment approach. The first recommendation is to optimize the first tier by introducing biotests and ecological indicators as screening tools, preferably *before* physical/chemical measurements are carried out. Ecotoxicological test systems in a first tier have the potential to detect the (synergistic) effects of complex mixtures of chemicals, including contaminants which are not routinely measured or even unknown toxic compounds. Also ecological indicators (e.g. presence or health of keystone species) may become helpful in selecting locations that require further risk assessment.

The second recommendation, regarding chemical measurements in the sediment, is not to measure only total contaminant concentrations, but also (or preferably) the bioavailable concentrations.

The third recommendation would be, when going from tier 1 to the next tiers, to ensure that:

- field surveys are included, in order to have an impact assessment;
- special attention should be given to exposure via food chains.
- bioassays are selected with sufficient ecological relevance
- cause-effect relationships are shown (resulting in lines of evidence with chemical data, bioassays and data from field surveys).

For tier 1 the focus may be different according to the objectives of the risk assessment:

- Assessment of potential hazard, taking into account mixture toxicity, forgotten substances etc;

- Ecological objectives in water management
- Management goals that have already been 'translated' into a list of causative substances that need to be addressed with special measures such as sediment remediation.

In case of the first option, tier 1 in Figure 3 would represent a preliminary assessment, consisting of a physical/chemical characterization of the sediment, "directed" by simple and sensitive (screening) bioassays. Effect-directed chemical characterization is expected to become a powerful and cost-effective approach when the mixture of chemicals is unknown. Ideally the bioassays of tier 1 are aimed at various toxicity mechanisms. Examples are bioassays that detect dioxin-like compounds or compounds with estrogenic effects (see [79]. The development of simple and sensitive tests that allow for a high throughput capacity is a very promising research field and will certainly gain interest in the future [16]. Screening bioassays may also be applied to obtain a first indication of the risks for the ecosystem. However, there still is a long way to go before these tests have been properly calibrated (with ecological data as reference) and consensus needs to be reached about the elements of such a ecotoxicological test system (screening battery) [80-82]. Screening bioassays can be performed on extracts of the sediments, but in order to complete the risk assessment, other tiers should then deal with the bioavailability of the contaminants.

In case ecological objectives in water management are the main reason to perform sediment risk assessment, the first focus should be on information from ecological surveys. This is expected to become more and more important with the implementation of the Water Framework Directive. The WFD will require information about the effectiveness of certain measures to improve the ecological quality of water bodies. Sediment remediation could be one of these measures, but proof has to be given that sediment quality is influencing the ecological quality of the water body. Ecological surveys may provide sensitive indicators that can be used as screening parameters in a first tier. The presence of keystone species, or measurements of their fitness [83] may be used in a screening step, in order to select locations that require further risk assessment.

There may, however, also be good reasons for starting tier 1 with physico-chemical analyses instead of biotests or ecological surveys. The presence of certain contaminants may already been known, or at least suspected based on information from pollution sources. When this is the case, or the risk assessment is simply started with a limited list of contaminants measured, simple first-tier risk assessment can be performed by calculating simple PEC/PNEC ratios. Also more advanced ecotoxicological models like OMEGA (species sensitivity distribution database) (see section 1) can be used, resulting in predictions of what groups of species might be affected and helping in the selection of test species for tier 2 and 3. Classical approaches will use total contaminant levels in the sediment, but the risk assessment can in this phase be much improved by including information on the bioavailable fraction of contaminants. This makes sense especially when the focus of the risk assessment is on known contaminants. Information on bioavailability will help to reduce the number of over- or underestimations of risk. However, this is not common practice yet.

In the case of remediation (environmental cleanup), *in situ* measures or no further action measures can be additional management options. Therefore, tier 2 of the *in situ* sediment risk assessment (see Figure 3) could start by checking a number of simple points:

- Are the contaminants present in the biologically relevant sediment layer (i.e. 0–10 cm of top-layer or even less deep) or only in deeper layers? Answering this question at this point requires data on the vertical sediment profile that are not always available, as sometimes tier 1 research is based on grab samples. However, it could be cost-effective to collect a minimum of information on the sediment profile and to use other sediment collection techniques that distinguish between the top-layer and sediment in deeper layers.

- Could there be natural attenuation because contaminated sediments will soon be capped by fresh sediments (in this case dredging would have no beneficial effect, see *Chapter 3 – Strategic Framework for Managing Sediment Risk at the Basin and Site-Specific Scale* and discussions in [84]).

In case the answers to these questions indicate that there might be (long-lasting) risks to the ecosystem, tier 2 then should be started by adding biological investigations to the assessment, in order to determine the condition of the ecosystem. The bioassays of tier 2 differ from those in tier 1, because their purpose is not to detect chemicals in a screening battery, but to provide evidence for adverse effects (that might be confirmed by field observations, so that a line of evidence emerges; see [5]). In tier 2 effect measurements include sublethal toxicity, such as reduced growth or impairment of reproduction. Biological observations in tier 2 may include 'classical' surveys of macrozoobenthos, bioaccumulation measurements and wildlife fitness parameters, while in tier 3 also more contaminant-specific endpoints like biomarker measurements, physiological or functional indicators [85,86], and molecular techniques can be used [87]. Sediments as a source for the bioaccumulation and biomagnification of contaminants is an important issue to deal with in tier 2 of the ecological risk assessment. Accumulation levels measured in organisms collected from the field can be used as proof of ecological risks, parallel to chemical analyses of bioavailable fractions.

If the role of the bioassays in tier 1 is to detect also non-priority compounds (as suggested above), bioassays with species that bear ecological relevance for the specific site will be needed in other tiers, in order to build lines of evidence (see above), or to demonstrate that there is a bioavailable fraction of toxic contaminants. This can be time-consuming and required only for a selection of sites that are studied further in tier 3. Also field bioassays can provide useful information and help to link laboratory results with ecological observations.

With regard to the selection of management options, there may be a need to base the decision on further research in a third tier (Figure 3). This research may comprise more location-specific biological, ecotoxicological and chemical research with the aim to detect what substances cause what effects and need to be managed. For this work, no general scheme can be given. Because the research is location-specific, it is tailor-made and countries are likely to differ in their approach. When the main driver for tier 3 is the EU WFD (failure to meet water quality targets), the research should focus on the relation between sediment quality and water quality. In that case, tier 3 should certainly include advanced chemical assessment techniques (e.g. availability). However, there may be local, regional or national management goals that also can be the driver for sediment remediation.

In Tier 3 also toxicity identification evaluation (TIE; [88]) work can be necessary to reveal the identity of toxic contaminants in the sediment (Fig. 3). TIE work may be similar to the effect directed chemical analysis in tier 1, but is likely to be focused more on the bioasssays chosen for tier 2 and 3, and to be used for the final steps in the risk assessment (e.g., demonstrating cause-effect relationships).

External pollution sources can significantly influence the success of any management remediation decision. Therefore identification, quantification and control (if necessary) of pollution sources has to be part of sediment management independently of whether sediments have to be dredged.

The results of tiers 2 and 3 could be integrated into an overall indicator of ecological quality. Ideally this is done by applying classification schemes that bear relevance for the link between observed effects in the field or in bioassays and the levels of contaminants in the sediment. A proposal for a terminology and a classification is given in Table 3.

Table 3: Proposed terminology and classification of in situ ecological risks caused by sediment contamination

Class	Risk caused by sediment contamination	Characteristics (proposal)	Reference: WFD classification for ecosystem quality[1]
5	Very high	Field observations indicate strong effects; bioassays confirm strong (acute) sediment toxicity[2]; contaminant levels (available fraction) exceed thresholds for serious effects by a factor of 10 or more and/or for effects on keystone-species[3].	Bad
4	High	Field observations indicate strong effects; bioassays confirm strong (acute) sediment toxicity[2]; contaminant levels (available fraction) exceed thresholds for serious effects.	Poor
3	Moderate	Field observations indicate significant, but moderate effects; bioassays confirm moderate (and long-term) sediment toxicity[2]; contaminant levels (available fraction) exceed no-effect thresholds.	Moderate
2	Minor	Field observations show no or minor effects; bioassays show no or moderate sediment toxicity[2]; contaminant levels (available fraction) exceed no-effect thresholds.	Good
1	None	Field observations show no effects; bioassays show no toxicity[2]; contaminant levels (available fraction) exceed no-effect thresholds.	High

[1] Causal relation always needs to be demonstrated when action is to take place solely for WFD targets.

[2] Toxicity can only be used in the sediment risk assessment if there is proof that sediment contamination is causing the effects in the bioassays. TIE procedures may be used to meet this criterion.

[3] On which the presence the functioning of the ecosystem is strongly dependent.

Human health risk

In the case of remediation (environmental cleanup), *in situ* measures or no further action measures can be additional management options. Therefore tier 2 of the *in situ* sediment risk assessment (see Figure 3) could start by checking a number of simple points:

- Are the contaminants present in a relevant sediment layer (i.e. 0–50 cm of top-layer or even less deep) or only in deeper layers?
- Could there be natural attenuation because contaminated sediments will soon be capped by fresh sediments (in this case dredging would have no beneficial effect).
- Is there reason for concern for effects on human health (are there fisheries active, is the location used as recreational site) ?

The next steps of the assessment of human health risks in tier 2 is to evaluate whether humans can be exposed at the contaminated location. Routes of exposure may include ingestion of contaminated river water, inhalation of chemicals volatilizing from sediment, recreational exposures (incidental ingestion of sediment and dermal contact with sediment and water), and ingestion of fish. Usually the primary risk is from ingestion of fish where chemical specific concentrations from sediment bioaccumulate. Exposure from fish consumption involves the determination of chemical concentrations in fish, the daily amount of fish ingested, the frequency of fish obtained from a specific source, and the duration of exposure. Risk characterization presents the calculated risks and hazards for the reasonably maximally exposed individual for each pathway and chemical, and across chemicals and pathways. A discussion on the uncertainties for all components of the risk assessment should also be included in the risk characterization. The goal is to provide this information in a manner that reflects transparency in the decision-making process, clarity in the communication, consistency with other assessments, and reasonableness. The risk characterization serves as the bridge between the assessment of potential risks from the site and the risk management decision concerning the potential need for remedial actions. Some elements of the risk assessment, such as pathogen and skin tests, may be regarded as optional and planned in a third tier (Figure 3).

A general scheme for assessment of human risk could be future work for the EC, because no such framework has yet been established.

Risk for groundwater contamination

In the case of remediation (environmental cleanup), *in situ* measures or no further action measures can be additional management options. Therefore, tier 2 of the *in situ* sediment risk assessment (see Figure 3) could start by checking a number of simple points:

- Are the contaminants present in deeper sediment layers (answering this question at this point requires data on the vertical sediment profile that are not always available, because sometimes tier 1 research is based on grab samples).
- Do the hydrological conditions give reason for concern for transport of contaminants to the groundwater?

Depending on the hydrological regime, risk assessment should focus on the possible transport of contaminants to and via the groundwater. The need for this is evident when there is a water flux from the sediment to aquifers. Also, in areas that are flooded, risk may be inferred (contaminants seeping through the soil to the aquifers). Most of the research should be planned in tier 2, but some elements of the risk assessment could be optional and planned in a third tier.

A general scheme for assessment of the risks for groundwater contamination cannot be given at this moment. The framework should be in line with the EU groundwater directive.

Risk for contamination of surface water
In the case of remediation (environmental cleanup), *in situ* measures or no further action measures can be additional management options. Therefore, tier 2 of the *in situ* sediment risk assessment (see Figure 3) could start by checking a number of simple points:

- Are the contaminants present in the relevant sediment layer (i.e. 0–10 cm of top-layer or even less deep) or only in deeper layers?
- Could there be natural attenuation because contaminated sediments will soon be capped by fresh sediments (in this case dredging would have no beneficial effect).
- Do the hydrological conditions give reason for concern for transport of contaminants to the surface water?
- Is the location an erosive site?

Contaminant remobilization can negatively affect surface water quality if the flux rates from the sediments via diffusion, advection, bioturbation or resuspension are higher than the water turnover, flow or mixing rates [84]. Most of the research should be planned in tier 2, but some elements of the risk assessment could be optional and planned in a third tier. A general scheme for the assessment of the risks of surface water (re)contamination by contaminated sediments cannot be given at this moment.

4.2. Risk assessment for DM management

The first tier for *ex situ* risk assessment (i.e., for managing the handling of dredged material) can be similar to that used for *in situ* sediment risk assessment, except that ecological observations will have to be replaced by models that predict the ecological risks (see Figure 3). It should be possible to decide on free relocation of the dredged material based on the results from tier 1 only (in case no effects are found or predicted). Also classification as resource (e.g. building material) or waste could be based on (primarily chemical) assessments in tier 1. Depending on this classification, the dredged material must be further investigated in tier 2, e.g. for waste classification or product requirements. The necessary steps in tier 1 and 2 will depend strongly on the existence of specific regulatory regimes that designate DM.

If necessary, the toxicity testing in tier 1 (for DM management), could be focused more on enlarging the window for chemical detection than on risk assessment. This can be done by using contaminant (group) specific receptor-based techniques, or by screening for certain modes of action, for example for mutagenicity. Those tests should be cost-effective (because they will be applied routinely) and be relevant to the risks of concern.

Tier 2 could be aimed at further risk assessment for sediments that fail the tests in tier 1, or at specific requirement for DM re-use options. Treatment and/or confined disposal of dredged material may be considered necessary on the basis of the outcome of tier 2.

While the sediments of major concern for *in situ* ERA are surface sediments to which benthic communities are exposed, there are a few further issues of concern if dredging is to be considered. First, contamination at depth might be at higher levels than it is at the surface, increasing *in situ* risk after dredging, so depth profiles for contamination should be considered.

In case a management option being considered for DM is free relocation (downstream) in the river, the effects on the sediment dynamics should be considered. For example, the risk assessment should cover high flow events and evaluate downstream sediment management issues caused by the relocation of dredged material. Possible negative effects on the quality of sediment settling in floodplains should be taken into consideration. Finally, consequences for irrigation and drinking water preparation might be included in the risk assessment.

The assessment of the effects of the dredging operation itself has received little attention so far. In some countries, environmental dredging is accompanied by a monitoring program, to ensure that at the location itself there are no or only temporarily negative side effects of the operation. The regulations of the birds and habitat directive should be mentioned in this respect.

5. Acknowledgements

All participants in SedNet working group 5 are thanked for their input in this paper. Mr P. Otte of the National Institute for Public Health and the Environment (RIVM) and J. Hugtenburg of the Institute for Inland Water Management and Waste Water Treatment, both in The Netherlands, are thanked for their contribution. Part of this chapter has also been published as a paper in the *Journal of Soils and Sediments* [26].

References

1. US EPA (1989): Risk Assessment Guidance for Superfund, Vol. 1. Human Health Evaluation Manual, EPA540/002, Washington DC.
2. Paustenbach DJ (1995): Retrospective on US Health Risk Assessment. Risk: Health, Safety & Environment 6, 283, Concord, New Hampshire.
3. Van Leeuwen CJ, Hermens JLM (1995): Risk assessment of chemicals: an introduction. Springer, New York.
4. US EPA (1998): Guidelines for Ecological Risk Assessment. EPA/630/R095/002F, Washington DC.
5. Burton GAJ, Batley GE, Chapman PM, Forbes VE, Smith EP, Reynoldson TB, Schlekat CE, Den Besten PJ, Bailer J, Green AS, Dwyer RL (2002): A Weight-of-Evidence Framework for Assessing Ecosystem Impairment: Improving Certainty in the Decision-Making Process. Human and Ecological Risk Assessment 8: 1675-1696
6. Hamelink JL, Landrum PF, Bergman HL, Benson WH (1994): Bioavailability: Physical, Chemical and Biological Interactions. CRC Press, Inc, Lewis Publishers: Boca Raton, FL.
7. Kraaij RH (2001): Sequestration and Biavailability of Hydrophobic Chemicals in Sediment. Thesis Institute for Risk Assessment Sciences Utrecht University, Utrecht, The Netherlands.
8. Hermens J, Canton H, Steyger N, Wegman R (1984): Joint effects of a mixture of 14 chemicals on mortality and inhibition of reproduction of Daphnia magna. Aquatic Toxicology 5: 315-322
9. Von Danwitz B (1992): Zur Abschätzung der Schadwirkung von Stoffkombinationen autf aquatische organismen (On the Estimation of Mixture Toxicity for Aquatic Organisms; in German), Thesis, University of Bremen, Bremen, Germany.
10. Lee BG, Griscom SB, Lee JS, Choi HJ, Koh CH, Luoma SN, Fisher NS (2000): Influences of Dietary Uptake and Reactive Sulfides on Metal Bioavailability from Aquatic Sediments. Science 287: 282-284
11. Lee JS, Lee BG, Luoma SN, Choi HJ, Koh CH, Brown CL (2000): Influence of Acid Volatile Sulfides and Metal Concentrations on Metal Partitioning in Contaminated Sediments. Environ Sci Technol 34: 4511-4516
12. Posthuma L, Suter GW, Traas TP (2002): Species Sensitivity Distributions in Ecotoxicology. CRC-Press: Pensacola, FL.
13. Ireland DS, Ho KT (2005): Toxicity tests for sediment quality assessments. In: Munawar M (Ed.), Ecotoxicological Testing of Marine and Freshwater Ecosystems, Vol. 1-42. CRC Press

14. Reynoldson TB, Bailey RC, Day KE, Norris RH (1995): Biological guidelines for freshwater sediment based on BEnthic Assessment of sediment (the BEAST) using a multivariate approach for predicting biological state. Australian journal of ecology 20: 198-219
15. Dirksen S, Boudewijn TJ, Slager LK, Mes RG, van Schaik MJM, De Voogt P (1995): Reduced Breeding Success of Cormorants in Relation to Persistent Organochlorine Pollution of Aquatic Habitats in the Netherlands. Environmental Pollution 88: 119-132
16. Den Besten PJ, Munawar M (2005): Synthesis and recommendations for ecotoxicological testing. In: Munawar M (Ed.), Ecotoxicological Testing of Marine and Freshwater Ecosystems: 249-260. CRC Press
17. Chapman PM (1996): Presentation and Interpretation of Sediment Quality Triad data. Ecotoxicology 5: 327-339
18. Den Besten PJ, Schmidt CA, Ohm M, Ruys MM, Van Berghem JW, Van de Guchte C (1995): Sediment Quality Assessment in the Delta of the Rivers Rhine and Meuse Based on Field Observations, Bioassays and Food Chain Implications. J Aquatic Ecosystem Health 4: 257-270
19. Babut MP, Ahlf W, Batley GE, Camusso M, De Deckere E, Den Besten PJ (2005): International Overview of Sediment Quality Guidelines and Their Uses. In: Moore DW (Ed.), Use of Sediment Quality Guidelines and Related Tools for the Assessment of Contaminated Sediments (SQG). Society of Environmental Toxicology and Chemistry (SETAC), Pensacola (FL)
20. Grothe DR, Dickson KL, Reed-Judkins DK (Eds.). 1996. Whole Effluent Toxicity Testing: An Evaluation of Methods and Prediction of Receiving Systems Impacts. SETAC Pellston Workshop on Whole Effluent Toxicity; September 16th-25th 1995; Pellston, MI. Pensacola FL: SETAC Pr
21. Tonkes M, De Graaf PJF, Graansma J (1999): Assessment of Complex Industrial Effluents in the Netherlands Using a Whole Effluent Toxicity (or WET) Approach. Wat Sci Tech 39(10-11): 55-61
22. Apitz S, White S (2003): A conceptual framework for river-basin-scale sediment management. JSS - J Soils & Sediments 3(3): 125-220
23. US EPA (1995): Guidance for risk characterisation. Science policy counsil, Washington DC.
24. Mozeto AA, Araújo PA, Kulmann ML, Silvério PF, Nascimento MRL, Almeida FV, Umbuzeiro GA, Jardim WF, Watanabe HM, Rodrigues PFao (2004): Integrated hierarchical sediment quality assessment program: Qualised project's approach proposal (São Paulo, Brazil); 2004. Presentation at SedNet meeting, January 29-31 2004, Lisbon, Portugal.
25. Magdaliniuk S, Pisano C, Vermersch M, Imbert T, Allard V (2000): Enlèvement des sédiments - Guide méthodologique - Evaluation détaillée des risques iés à la gestion des sédiments et aux opérations de curage (santé humaine, ressources en eau). Prepared by Pôle de compétence sur les sites et sols pollués - Agence de l'Eau Artois-Picardie, DIREN Nord-Pas de Calais Rep. Nr.: 31/5/2000.
26. Den Besten PJ, de Deckere E, Babut M, Power B, DelValls A, Zago C, Oen A, Heise S (2003): Biological Effects-based Sediment Quality in Ecological Risk Assessment for European Waters. J Soils & Sediments 3(3): 144-162
27. Swartjes FA (1999): Risk-Based Assessment of Soil and Groundwater Quality in the Netherlands: Standards and Remediation Urgency. Risk Analysis 19: 1235-1249
28. Van Elswijk M, Hin JA, Den Besten PJ, Van der Heijdt LM, Van der Hout M, Schmidt C, A (2001): Guidance Document for Site-Specific Effect-Based Sediment Quality Assessment. AKWA Report 01.005 / RIZA Rapport 2001.052, Institute for Inland Water Management and Waste Water Treatment (RIZA), Lelystad, The Netherlands (in Dutch).

29. Otte PF, Van Elswijk M, Bleijenberg M, Swartjes F, Van der Guchte C (2000): Calculation of criteria for human risk assessment for contaminated sediment. RIZA internal report 2000.084x, Institute for Inland Water Management and Waste Water Treatment (RIZA), Lelystad, The Netherlands (in Dutch).

30. Long E, Chapman PM (1985): A Sediment Quality Triad: Measures of Sediment Contamination, Toxicity, and Infaunal Community Composition in Puget Sound. Marine Pollution Bull 16(10): 405-415

31. De Pauw N, Heylen S (2001): Biotic index for sediment quality assessment of watercourses in Flanders, Belgium. Aquatic Ecology 35: 121-133

32. De Deckere EMGT, De Cooman W, Florus M, Devroede-Vanderlinden M-P (2000): Handboek voor de karakterisatie van de bodems van de Vlaamse waterlopen, volgens de Triade, 2de versie (Guidance Document for Characterisation of Flemish Waters According to the Triad, 2nd version - in Dutch). Ministerie van de Vlaamse Gemeenschap, AMINAL/ afdeling Water, Brussels.

33. Van der Zandt PTJ, Van Leeuwen DJ (1992): A proposal for priority setting of existing chemical substances. Studie i.o.v. EC, Dir-Gen voor Milieu, kernenergie en civiele bescherming.

34. Garric J, Bonnet C, Bray M, Migeon B, Mons R, Vollat B (1998): Bioessais sur sédiments: Méthodologie et application à la mesure de la toxicité de sédiments naturels. 97.9004. Agence de l'Eau Rhône-Mediterranée-Corse.

35. Garric J, Flammarion P, Bonnard R, Roger MC (2002): Etude de l'impact de la contamination toxique du sédiment sur les biocénoses benthiques: Mise en relation micropolluants - Liste faunistique (IBGN). Agence de l'Eau Rhône-Méditérannée-Corse.

36. Imbert T, Py C, Duchene M (1998): Enlèvement des sédiments - Guide méthodologique - Faut-il curer ? Pour une aide à la prise de décision. Pôle de compétence sur les sites & sols pollués Nord-Pas de Calais - Agence de l'Eau Artois-Picardie, Douai.

37. Müller G (1979): Schwermetalle in den Sedimenten des Rheins - -Veränderungen seit 1971. Umschau 79: 779-783

38. ATV-DVWK (2003): Umgang mit Baggergut. Teil 2: Fallstudie. Entwurf. Deutsche Vereinigung für Wasserwirtschaft. 80 pp.

39. Federal Environment Agency (2001): Environmental Policy - Water Resources Management in Germany. Part II - Quality of Inland Surface Waters, Federal Ministry for the Environment, Nature, Conservation and Nuclear Safety, Division WAI 1B.

40. Ahlf W, Hollert H, Neumann-Hensel H, Ricking M (2002): A Guidance for the Assessment and Evaluation of Sediment Quality: A German Approach Based on Ecotoxicological and Chemical Measurements. J Soils & Sediments 2(1): 37-42

41. Henschel T, Maaß V, Ahlf W, Krebs F, Calmano W (2001): Gefährdungsabschätzung von Gewässer-sedimenten - Handlungsempfehlungen und Bewertungsvorschläge. Teil III. Document für den Fachausschuss Gewässersedimente. Untersuchung und Bewertung von Sedimenten - Ökotoxikologische und chemische Testmethoden. Springer Verlag, Berlin, S. 493-496.

42. D.Lgs (1999): 'Disposizioni sulla tutela delle acque dall'inquinamento e recepimento della direttiva 91/271/CEE concernente il trattamento delle acque reflue urbane e della direttiva 91/676/CEE relativa alla protezione delle acque dall'inquinamento provocato dai nitrati provenienti da fonti agricole', published in G.U. 29 May 1999, n. 124.

43. Van der Gaag MA, Stortelder PBM, Van der Kooij LA, Bruggeman WA (1991): Setting Environmental Quality Criteria for Water and Sediment in The Netherlands: a Pragmatic Ecotoxicological Approach. Europ Water Poll Control 1: 13-20

44. Van der Kooij LA, Van de Meent D, Van Leeuwen CJ, Bruggeman WA (1991): Deriving Quality Criteria for Water and Sediment from the Results of Aquatic Toxicity Tests and

Product Standards: Application of the Equilibrium Partitioning. Method. Water Res 25: 697-705

45. Van de Guchte C, Beek M, Tuinstra J, Rossenberg M (2000): Quality Guidelines for the Management of Water Systems. Commission for Integral Water Management, The Hague (in Dutch).

46. CUWVO (1990): Recommendations for the monitoring of compounds of the M-list of the national policy document on water management. Water in the Netherlands: A time for action. Commission for the Implementation of the Act on Pollution of Surface Waters (CUWVO), The Hague, Netherlands (in Dutch).

47. Saaty TL (1980): The Analytic Hierarchy Process. Mc Graw-Hill: New York.

48. Mesman M, Rutgers M, Peijnenburg WJGM, Bogte JJ, Dirven-Van Breemen ME, De Zwart D, Posthuma L, Schouten AJ (2003): Site-specific ecological risk assessment: the Triad approach in practice. Conference Proceedings ConSoil. 8th International FZK/TNO conference on contaminated soil, pp. 649-656.

49. Cornelissen G, Rigterink H, Ten Hulscher TEM, Vrind BA, Van Noort PCM (2001): A simple Tenax extraction method to determine the availability of sediment-sorbed organic compounds. Environmental Toxicology and Chemistry 20: 706-711

50. SFT (1997): Classification of Environmental Quality in Fjords and Coastal Waters (in Norwegian). SFT Report 97:03. Norwegian Pollution Control Authority, Oslo.

51. SFT (1997): Classification of Environmental Quality in Freshwater (in Norwegian). SFT Report 97:04. Norwegian Pollution Control Authority, Oslo.

52. NRA (1995): Risk Assessment of Contaminated Sediment. Prepared by WRc plc. R&D Note. Prepared by Fleming R et al.

53. SNIFFER - Scotland and Northern Ireland Forum for Environmental Research (1995): Freshwater Sediment Assessment - Scoping Study. Prepared by WRc plc, Report No. SR 3931/1 for SNIFFER. Prepared by Fleming, R, Johnson I, Delaney P, Reynolds, P.

54. Marine Pollution Monitoring Management Group (1998): National Monitoring Programme: Survey of the Quality of UK Coastal Waters.

55. Environment Agency of England and Wales (1997): EC Dangerous Substances Directive (76/464/EEC), Monitoring Requirements 11 pages. December 6th, 1996.

56. Environment Agency of England and Wales (2000): Proposed Environmental Quality Guidelines for Dioxins and Furans in Water and Sediments. Prepared for the Environment Agency by WRc plc, Grimwood, M J, Mascarenhas R, Sutton A. R&D Technical Report P48.

57. Environment Agency of England and Wales (2002): Review and Recommendations of Methodolgoies for the Derivation of Sediment Quality Guidelines. R&D Technical Report P2-092/TR. Prepared by CEFAS for the Environment Agency of England and Wales. August 2002.

58. Environment Agency of England and Wales (2002): Scoping Study - Sediments in England and Wales: Nature and Extent of the Issues. Prepared by Beth Power for the Environment Agency of England and Wales. February 2002.

59. Apitz SE, Crane M, Power EA (2004): Use of Sediment Quality Values (SQVs) in the Assessment of Sediment Quality. Draft report to the Environment Agency of England and Wales.

60. Chapman PM (2000): The Sediment Quality Triad: then, now and tomorrow. Int J Environ Pollution 13: 351-356

61. DelValls TA, Chapman PM (1998): Site-specific Sediment Quality Values for the Gulf of Cádiz (Spain) and San Francisco Bay (U.S.A), Using the Sediment Quality Triad and Multivariate Analysis. Ciencias Marinas 24: 313-336

62. Riba I, DelValls TA, Forja JM, Gómez-Parra A (2002): Influence of the Aznalcóllar Mining Spill on the Vertical Distribution of Heavy Metals in Sediments from the Guadalquivir Estuary (SW Spain). Marine Pollution Bull 44: 39-47

63. Riba I (2003): Evaluación de la calidad ambiental de sedimentos de estuarios afectados por actividades mineras mediante métodos integrados. Tesis Doctoral. Universidad de Cadiz.

64. DelValls TA, Forja JM, Gómez-Parra A (2002): Seasonality of Contamination, Toxicity, and Quality Values in Sediments from Littoral Ecosystems in the Gulf of Cádiz (SW Spain). Chemosphere 7: 1033-1043

65. Apitz SE, Ayers B, Kirtay VJ (2004): The Use of Data on Contaminant/Sediment Interactions to Streamline Sediment Assessment and Management, Final Report, Prepared for Y0817 Navy Pollution Abatement Ashore Technology Demonstration/Validation Program. San Diego, CA: SPAWAR Systems Center San Diego. Report nr 1918.

66. National Research Council (NRC) (1997): Contaminated Sediments in Ports and Waterways: Cleanup Strategies and Technologies. National Academy Press, Washington, D.C.

67. Babut MP, Perrodin Y (2001): Evaluation écotoxicologique de sédiments contaminés ou de matériaux de dragage. (I) Présentation et justification de la démarche. Rapport d'Etudes. Voies Navigables de France (VNF)-Centre d'Etudes Techniques Maritimes et Fluviales (CETMEF), 47 pp. (http://www.lyon.cemagref.fr/bea/tox/dragage.html).

68. Hagner C, Peters C (2001): The European and international policy framework. In: Salomons W (Ed.), Dredged Material in the Port of Rotterdam - Interface between Rhine Catchment Area and North Sea. Part D: 138-159. GKSS Research Centre, Geesthacht, Germany

69. Bundesanstalt für Gewässerkunde B (2000): Handlungsanweisung für den Umgang mit Baggergut im Binnenland (HABAB-WSV). pp.

70. Bundesanstalt für Gewässerkunde B (1999): Handlungsanweisung für den Umgang mit Baggergut im Küstenbereich (HABAK-WSV). pp.

71. Krebs F (1988): Der pT-Wert: Ein gewässertoxikologischer Klassifizierungsmaßstab. GIT Fachzeitschrift für das Laboratorium 32: 293-296

72. Krebs F (2005): The pT-method as a Hazard Assessment Scheme for sediments and dredged material. In: Férard J-F (Ed.), Small-scale Freshwater Toxicity Investigations, Vol. 2 (Hazard Assessment Schemes): 281-304. Springer, Dordrecht, The Netherlands

73. Ministry of the Environment VWA, President of Veneto Region, Mayor of Venice, Mayor of Chioggia, President of Venice Province (1993): Protocol 8/4/93 for the Classification of dredged material, Venice, Italy.

74. Stronkhorst J (2003): An Effect-based Assessment Framework to Regulate the Disposal of Contaminated Harbour Sediments in Dutch Coastal Waters. PhD Thesis, Free University, Amsterdam.

75. Schipper CA (2004): Implemetation of CTT test. Report of National Institute for Coastal and Marine Management (RIKZ), The Hague.

76. Schipper CA, Klamer JC (2006): Evaluation of CTT test. Report of National Institute for Coastal and Marine Management (RIKZ), The Hague.

77. MD (1997): Regulation. The Control of Dredging and Dumping in the Sea and Watercourses (in Norwegian). Fastsatt av Miljøverndepartementet 4. desember 1997.

78. Bates AD, Hooper AG (1997): Inland dredging - Guidance on good practice. CIRIA Publication Code: R169. 184 pp. ISBN: 0 86017 477 8.

79. Van der Burg B, Brouwer A (2005): Bioassays and Biosensors: Capturing Biology in a Nutshell. In: Munawar M (Ed.), Ecotoxicological Testing of Marine and Freshwater Ecosystems: 177-194. CRC Press

80. Ahlf W, Braunbeck T, Heise S, Hollert H (2002): Sediment and Soil Quality Criteria. In: Günther A (Ed.), Environmental Monitoring Handbook: 17.11 - 17.18. McGraw-Hill, New York

81. Ahlf W, Heise S (2005): Sediment toxicity assessment: Rationale for effect classes. JSS - J Soils & Sediments 5(1): 16-20

82. Heise S, Ahlf W (2002): The need for new concepts in risk management of sediments. J Soils & Sediments 2(1): 4-8

83. Rowe CL, Hopkins WA, Congdon JD (2001): Integrating individual-based indices of contaminant effects: How multiple sublethal effects may ultimately reduce amphibian recruitment from a contaminated breeding site. The Scientific World 1: 703-712

84. Apitz SE, Davis JW, Finkelstein K, Hohreiter DW, Hoke R, Jensen RH, Jersak J, Kirtay VJ, Mack EE, Magar VS, Moore D, Reible D, Stahl (Jr.) RG (2005): Assessing and Managing Contaminated Sediments: Part II, Evaluating Risk and Monitoring Sediment Remedy Effectiveness . Integrated Environmental Assessment and Management (online-only) 1(1): e1–e14

85. Griebler C (1996): Some applications for the DMSO-reduction method as a new tool to determine the microbial activity in water-saturated sediments. Arch.Hydrobiol.Suppl. 113(Large Rivers 10): 405-110

86. Leipe T, Kersten M, Pohl C, Heise S, Witt G, Liehr G (2005): Environmental Geochemistry of a Historical Dumping Site

87. in the Western Baltic Sea. Marine Pollution Bulletin 50: 446-459

88. Lange A, Maras M, De Coen WM (2005): Molecular methods for gene expression analysis: ecotoxicological applications. In: Munawar M (Ed.), Ecotoxicological Testing of Marine and Freshwater Ecosystems: 153-176. CRC Press

89. Ankley GT, Schubauer-Bergian MK (1995): Background and overview of current sediment toxicity identification evaluation procedures. J. Aquatic Ecosystem Health 4: 133-149

Sediment regulations and monitoring programmes in Europe

Helge Bergmann [a] and Vera Maass [b]

[a] *Federal Institute of Hydrology, Am Mainzer Tor 1, D-56068 Koblenz, Germany*
[b] *Hamburg Port Authority, Dalmannstraße 1, D- 20457 Hamburg, Germany*

1. Introduction

This chapter gives an overview of existing sediment regulations and monitoring activities in Europe. In addition, several comments by the authors on the present situation and future development are presented.

Water bodies such as rivers and lakes serve society in many ways, as:
- a natural collector for rain, snow and groundwater
- a natural ecosystem or even a nature protection area
- a drinking-water reservoir
- a reservoir for irrigation
- a recipient of and free transport system for waste
- a recreation area for water sports
- an area for commercial or sport fishing
- and, finally, a waterway for navigation.

From this list it may be clear that there will be conflicting interests among the different stakeholders with regard to the desired functions of a river, lake or estuary. This also means that sediment management will be confronted with the interests and claims of several different social or economic groups. As a consequence, it is necessary to agree on the functions desired and to set up rules and regulations to optimise the management of sediments within a water body on the regional, national or transboundary level.

Management of sediments, in particular of dredged material, falls into many domains: engineering, environmental protection, ecology, economy and legislation, some of them having conflicting objectives, as addressed in chapter 2 – *Sediment Management Objectives and Risk Indicators*. Therefore, for optimal sediment management with a minimum of ecological impact and cost, it seems useful to integrate demands and boundary conditions from these fields into suitable and practical regulations.

An integral part of sediment management is monitoring of aquatic sediment, which provides the data necessary to assess potential risks evolving from (contaminated) sediments, to control the effects of measures taken to improve a situation and to check compliance with agreed objectives.

The view presented in this chapter focuses on political and administrative perspectives, whereas scientific approaches and tools are described in several preceding chapters of this book.

2. Regulatory situation

2.1. Definition of sediments in environmental management

The lithosphere globally comprises the solid constituents of the planet earth. In scientific and political domains, this solid phase is referred to, for example, as earth, soil, sediment, rock, sand, clay, dredged material or waste, depending on the respective scientific frame, the political regulatory framework or the management objective. International definitions describe soil and sediment as being of natural origin (Table 1).

Table 1: Some international definitions of soil, sediment and dredged material

Soil ISO [1]	"Upper layer of earth's crust composed of mineral parts, organic substance, water, air, and living matter."
Sediment PIANC [2]	"Material, such as sand, silt, or clay, suspended in or settled on the bottom of a water body. Sediment input to a body of water comes from natural sources ... or as the result of anthropogenic activities..."
SedNet [3]	"Sediment is an essential, integral and dynamic part of our river basins."
WFD AMPS [4]	"Sediment is particulate matter such as sand, silt, clay or organic matter that has been deposited on the bottom of a water body and is susceptible to being transported by water"
Dredged material OSPAR Convention [5]	"2.3 In the context of these guidelines, dredged materials are deemed to be sediments or rocks with associated water, organic matter etc. removed from areas that are normally or regularly covered by water."
London Convention [6]	"1.1 .....The greater proportion of the total amount of material dredged world-wide is, by nature, similar to undisturbed sediments in inland and coastal waters..."
[7]	"The term 'dredged material' refers to material that has been dredged from a water body, while the term sediment refers to material in a water body prior to the dredging process."
[8]	"Materials excavated during e.g. maintenance, construction, reconstruction and extension measures from waters. Note: Dredged material may consist of: Sediments or subhydric soils, soils and parent material beneath the surface water body."

2.2. Regulatory instruments

The management of aquatic sediments includes amongst others the following activities:
- Capital (= new) and maintenance dredging
- Beneficial use of the material dredged
- Relocation or disposal of dredged material in water or on land (unconfined, confined)
- Treatment, capping, other technical options
- Remediation of contaminated sites
- Mineral extraction

The main instruments for managing aquatic sediments and dredged material that are available and applied in EU countries vary in their legal status:

-	Laws		obligatory
-	Directives		obligatory
-	Decrees, ordinances		obligatory
-	Guidelines	- international	mostly not obligatory
		- national	mostly obligatory
-	Concepts		mostly not obligatory
-	Action values, guide values		regionally derived, obligatory

2.3. United Nations Economic Commission for Europe

Under the umbrella of the United Nations Economic Commission for Europe (UNECE), several conventions have been designed in the transboundary context. One of them is the "Convention on the Protection and Use of Transboundary Watercourses and International Lakes" signed in Helsinki in 1992 [9].

Among its objectives are transboundary cooperation and assistance among countries which are setting up or renewing joint bodies (e.g. river and lake commissions, border commissions). Cooperation is strengthened between the UNECE conventions related to transboundary waters, industrial accidents, environmental impact assessment, public participation and long-range transboundary air pollution. It is also intended to assess in joint groups experience gained with international legal instruments on the protection and use of waters. Difficulties encountered when implementing the Convention, arising from differences in administrative practice, management and protection responsibilities or water use rights in riparian countries will also be examined. Specific tasks undertaken by joint bodies under the Convention include:
- Identification of pollution sources, inventories and exchange of information on pollution sources, joint monitoring programmes, warning and alarm procedures, emission limits for waste water;

- Evaluation of the effectiveness of control programmes;
- Implementation of environmental impact assessment.

The major objective of the Convention lies in promoting cooperation among riparian states in a water catchment area, and in developing generic guidance for transboundary water management. In this context, the Convention also deals with monitoring sediments, including the development of guidelines for sampling and analysis. However, it is not intended to issue legally binding decisions or to carry out joint monitoring programmes.

2.4. European Union

EU legislation includes a number of directives directly targeting sediment or soil, some of which also directly or indirectly address dredged material [10]:

- The **European Framework Directive on Waste** [11] is implemented in the national waste regulations of the member states. Some of the European and national waste regulations are relevant to dredged material.
- In the **European Waste Catalogue** [12] dredged material is mentioned in the waste code 17 05 (Soil and dredging spoil), differentiated in sub-codes 17 05 05 (Dredging spoil containing dangerous substances) and 17 05 06 (Dredging spoil other than those mentioned in 17 05 05). The classification in these sub-codes is the determination of non-hazardous or hazardous waste, which is of significant cost-relevance in practice. If dredged material is waste, usually it is non-hazardous according to the chemical criteria in the European Waste Catalogue.
- The **European Landfill Directive** [13] regulates waste landfilling, including dredged material. If dredged material is to be disposed of upland with no further use in view, the appropriate landfill category has to be determined. Technical demands for those landfill categories (disposal facilities) are defined in the European Landfill Directive.
- The **Water Framework Directive** (WFD) [14] is a comprehensive directive of the European Parliament and of the Council establishing a framework for community action in the field of water policy. Although ecologically oriented, the keyword "sediment" appears only eight times (s. Annex 1 at the end of this chapter)

Beyond the directives mentioned above, there are a number of Council Directives that do not specifically address sediment or dredged material but which may still relate to sediment management. They are summarised in Table 2. More detailed descriptions are given in chapter 2 – *Sediment Management Objectives and Risk Indicators*.

A European **soil framework directive** does not yet exist. However, in 2002 the European Commission issued a communication paper as a basis for discussion on the development of an **EU Strategy for Soil Protection** [15, 16, 17]. Five technical working groups were assembled to help develop the Thematic Strategy. Working group final reports were published in April 2004 [18]. The European Commission is expected to

Table 2: European Council Directives potentially concerning sediment management

Directive	Document
Environmental Impact Assessment	97/11/EC
Surface Water Directive	75/440/EEC
Drinking Water Directive	80/778/EEC
Bathing Water Quality Directive	76/160/EEC
Groundwater Directive	80/68/EEC
Groundwater Daughter Directive	COM(2003)550
Conservation of Natural Habitats and of Wild Fauna and Flora Directive	92/43/EEC
Fish Water Directive	78/659/EEC
Shellfish-Water Directive	79/923/EEC
Wild Birds Directive	79/409/EEC
Council Directive on Waste	75/442/EEC
Dangerous Substances Directive	76/464/EEC
Urban Wastewater Treatment Directive	91/271/EEC

come forward in 2006 with a policy package containing a Communication assessing the current situation and outlining the strategy's objectives, a detailed Extended Impact Assessment and a Soil Framework Directive addressing the key soil threats outlined in the Commission's original Communication [19].

With the **"Marine Strategy"** another EU policy framework was developed with the objective of updating the European maritime policy. This issue and six further Thematic Strategies were conceived as a new way of approaching environment policy, looking at themes in a holistic way and emphasising integration of environment in other policies and programmes as the main route to achieving environmental aims. The Marine Strategy, being part of the project, is aimed at protecting Europe's seas and oceans and ensuring that human activities in these seas and oceans are carried out in a sustainable manner.

At present this policy is laid down in proposals for a Marine Strategy Directive [20] and an Impact Assessment Framework [21]. In which way sediments and sediment management will be addressed is still open.

Implementation of all of these directives is obligatory for EU member states, unlike many other decisions and recommendations of other international bodies.

2.5. International maritime conventions

Sediment management in coastal areas often involves dredged material management, i.e. the more or less regular dredging of large volumes of sediment in estuaries and its disposal either in the aquatic environment or on land. As part of this sediment is contaminated, the international maritime conventions to which most European countries belong have developed specific guidelines for the input or relocation of this sediment in convention waters. Table 3 gives an overview of the existing guidelines on existing regulations.

Table 3: Sediment (dredged material) regulations of international maritime conventions in Europe and worldwide

Convention name	Convention area	Regulation (year)	Reference
LONDON	Globally marine areas	Specific guidelines for assessment of dredged material (2000)	[6]
OSPAR	NE Atlantic, North Sea	Revised OSPAR Guidelines for the Management of Dredged Material (2004)	[5]
HELSINKI	Baltic Sea	Dredged Material Guidelines (1994)	[22]

For decision-making in sediment management, a relative quality scale with defined sediment quality criteria (SQC) has to be developed. This quality indicator, called action value, guide value or limit value, functions as a yardstick to show decision-makers how the sediment should or should not be managed. Both the London and the OSPAR dredged material guidelines call for the development of such "action values".

In a number of countries in Europe and around the world, lists with such SQCs for **contaminant concentrations** are in use, often in the form of lower and upper action values [23]. These values were not developed in a standardised way but on the basis of the respective scientific considerations and national objectives. Therefore, they vary considerably from country to country and are far from offering a firm basis for decision making. For this reason, scientists warn against a simplified application of these action values as the only basis for "pass-fail" decisions in dredged material management [24].

Complementing the assessment of contaminant concentrations in sediments, ecotoxicological assays, investigating the **effects on biota,** have been developed. The rationale for this complementary instrument was that it is not possible to analyse the huge number of hazardous chemicals potentially present in sediments. Further, it is scientifically impossible to derive chemical quality criteria for their assessment. This dilemma can be partly resolved by application of bioassays integrating adverse effects exerted by the hazardous substances in a sediment sample (see also chapter 5 – *Risk Assessment Approaches in European Countries* and the SedNet Book on "Sediment quality and impact assessment of pollutants" by D. Barceló & M. Petrovic (Eds.), this series).

Ecotoxicological sediment quality criteria similar to those for concentrations have been developed in only a few countries and are not standardised.

2.6. International commissions for transboundary rivers

Many European countries successfully co-operate in the environmental management of rivers. An overview of the major river basins is listed in Annex 2. Generally, their strategies and action plans focus on pollution abatement and improvement of the ecological status of the waters. However, with one exception, none of them have developed guidelines with respect to the management of sediment or dredged material. Obviously, among the main components of a water body - water, sediment, biota - the sediment phase does not yet play a major role in river management. Only the International Commission for the Protection of the river Rhine (ICPR) has established an agreement on national requirements for the disposal and relocation of dredged material [25].

This situation is not in line with the role sediments play as an integral component in any water body, hence in environmental management, ranging from hydrology (e.g. erosion, sedimentation) through chemistry (contamination sink and source) to ecology (natural living space for biota on and in sediments). This situation is mirrored by the European Water Framework Directive which does not pay much specific attention to sediments.

2.7. National regulations

National regulations for sediment management deal primarily with dredged material. The largest part of it is handled in coastal areas due to natural hydrological processes (input into the OSPAR Convention area ca. 100 Mio. m^3 annually). Therefore, it is not surprising that this issue has found much more attention from the political and engineering concerns than inland dredging. As a consequence, there are a number of national regulations concerning coastal sediments and only a few for inland sediments.

Some OSPAR Contracting Parties have invested significant resources in developing their national guidelines and action levels for dredged material in coastal areas, and others are in the process of developing or reviewing them. A report has been compiled on the Contracting Parties' situation, including a survey of the approaches used to derive their values [23]. However, there was no wide support for establishing a *common* OSPAR approach on national action levels.

For sediment management in freshwater areas, it is difficult to form a comprehensive picture of existing European regulations or guidelines, because information on sub-basin management is hard to obtain.

For example, among the sixteen federal states of Germany, there is a high degree of co-operation in water management, but no joint regulation for dredged material management exists.

2.8. Specific dredged material issues

Dredged material and waste

In a number of EU directives, dredged sediment is categorised as waste. However, other international definitions do not see dredged material *a priori* as waste. This raises the question of where the term "waste" enters the picture and what generally marks "dredged material" from a legislator's viewpoint as "waste". Or, considering a scale from clean natural sediment to heavily contaminated material, at what point does "dredged material" become "waste"?

An explanation may be found in a directive of the European Union, the Framework Directive on Waste [11], which states in Article 1:

- "For the purposes of this Directive 'waste' shall mean any substance or object in the categories set out in Annex I which the holder discards or intends or is required to discard."
- The corresponding Annex I (Categories of Waste) defines under category Q16:
  "Any materials, substances or products which are not contained in the above categories."

The collective category Q16 includes also dredged material. This means that simply the intention or the requirement to discard sediment after dredging turns this material into waste regardless of its degree of contamination. This is a theoretical view, far away from any hydrological and ecological considerations. It disregards the natural sources and the role of sediments in the aquatic environment, from water bed stability to acting as substrate for biota.

The fact is that dredged material may be, and in reality often is, mainly clean sediment not exceeding agreed sediment quality criteria. Such material is still designated and has to be handled as waste. If strictly applied, this definition would prevent large volumes of clean sediments from being relocated in watercourses despite the minimal ecological impact and costs.

In addition, this definition is not in line with the spirit of the "polluter pays" principle established in European legislation [11] in so far as the "polluter" is released from his legal obligation and the dredging authority is made liable for the economic consequences (see section *Control of Contaminant Sources* below).

On the other hand, dredged material may be contaminated to an extent rendering it environmentally unacceptable for unrestricted use. In that case the European Directive on Hazardous Waste (91/689/EEC) is applicable, defining "hazardous waste":

> **Annex I.B** - Wastes which contain any of the constituents listed in *Annex II* and having any of the properties listed in *Annex III* and consisting of:
> - Category 23. soil, sand, clay **including dredging spoils**;
>
> **Annex II** - Constituents of the wastes in Annex I.B which render them hazardous when they have the properties described in Annex III
> - Wastes having as constituents: [List of contaminants]
>
> **Annex III** - Properties of wastes which render them hazardous:
> - Property H14 'Ecotoxic': substances and preparations which present or may present immediate or delayed risks for one or more sectors of the environment.

In that case, special care must be taken in handling this material, and the potential impacts of disposal must be assessed according to the international guidelines for the management of dredged material. The graph below visualises the range of dredged material quality between "clean/harmless" and "contaminated/harmful":

Due to the natural origin of sediments (in contrast to man-made waste), their characteristics and their ecological role in water bodies, it seems appropriate to consider their re-definition in the context of waste. For the same reason, the development of specific guidance for the management of dredged material has been proposed in the Final SedNet Conference [26].

Control of contaminant sources

The term "control of contaminants at the source" defines a political concept in European environmental policy which holds that the producer of waste is responsible for its legal handling and for the costs of disposal or treatment. In this way, the traditional "costs-by-cause" principle has been applied to environmental contamination, where it is called the "polluter pays" principle. Its basic intention is described in the European Framework Directive on Waste [11]:

"In accordance with the 'polluter pays' principle, the cost of disposing of waste... shall be borne by:
- the holder who has waste handled by a waste collector or by an undertaking...
- and/or the previous holders or the producer of the product from which the waste came."

As far as sediments and specifically the disposal of dredged material is concerned, this principle is hardly applied in real projects. If dredged material is found to be contaminated, in most cases it has been contaminated by a source outside the control range of the dredging management. However, experience shows that the contaminant source (the "polluter") is usually neglected when seeking options for handling the dredged material. Contrary to the above principle, the (primary) polluter is not held liable for the costs. For many years this attitude, "be strict on dredged material, but neglect contamination sources" has not been adequately taken into account in regulatory frameworks with regard to sediments and dredged material.

Part of the problem lies in the insufficient and incoherent handling of the great number of contaminant sources in a *river basin*. A single source may be under control, fulfilling even the legal obligations of a *waste water* regulation. However, it is the sum of the single contaminant emissions which accumulates in the sediment and, hence, in the dredged material downstream. Apparently, the requirements for emission standards do not include a standard like "not leading to excess or harmful accumulation in sediment". Such a requirement can only be implemented in a basin-wide management approach. (A similar concept was developed for global air pollution, which finally led to "pollution shares" whose *sum* was not to exceed agreed limits.)

Progress in improving this situation in Europe can be expected by two new directives concerning this field:

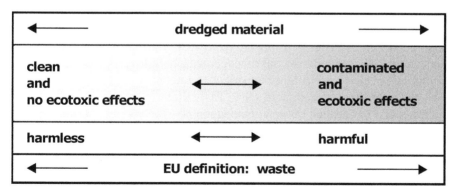

1. The Water Framework Directive [14]. It calls for river basin-scale water management, in which upstream and downstream situations, sources and

sinks of contaminants are combined in a comprehensive river management plan.

2. The Directive on environmental liability with regard to the prevention and remedy of environmental damage [27]. Its objective is the implementation of the prevention and remedying of environmental damage through the furtherance of the "polluter pays" principle. The fundamental principle will therefore be that an operator whose activity has caused the environmental damage is to be held financially liable.

It will be beneficial to the aquatic environment when the "polluter pays" principle becomes a firm political objective within sediment management. Including it in the effective control of contaminants at the source in *all* concepts and regulations will be a significant improvement, because:

- Downstream authorities responsible for handling sediments and dredged material will no longer be obliged to remove contaminated sediments without having influence over external (upstream) contaminant sources.
- The emitter responsible for the contamination can be urged to accept a share of the costs of the management solution.

2.9. Examples of Co-operation

Dutch–German Exchange on Dredged Material (DGE)
Economically efficient and environmentally sound management and handling of dredged material is important both in The Netherlands and in Germany, as huge amounts of dredged material emerge from maintenance, construction and remedial (clean-up) works within water systems. The national volumes of dredged material amount to roughly 35 and 50 million m³/year, respectively.

Against this background and because of progressing European development, in 1999 the competent governmental authorities in The Netherlands and Germany started a Dutch–German Exchange on dredged material (DGE) comprising stakeholders from ministries, scientific institutions, navigation administrations and port authorities. The main objective of this working group is to increase understanding of dredged material management, both in terms of policy and on the practical level, by exchanging experiences and knowledge. In addition, DGE seeks to contribute throughout the EU, directly or indirectly, to the standardisation of dredged material management – for example, in the European Sediment Network (SedNet).

From this perspective, DGE covers different aspects of dredged material management such as legislation, treatment, chemistry, ecotoxicology, dredging technology, etc. Sharing knowledge in these fields requires an analysis (classification and comparison) of the dredged material terminology of both countries in order to reach a common understanding of the important terms, despite the legal or technical differences and difficulties with regard to the correct translation of terminology used in Germany and The Netherlands. Information on different aspects is presented in special DGE publications [28].

European Sediment Networking (SedNet)
As described earlier, SedNet is a European sediment research network for developing environmentally, socially and economically viable practices for sediment management on the river basin scale [3]. Given its transboundary nature, no single water manager or country is responsible for dealing with sediment management problems at this scale. SedNet was established to help structure and facilitate a European approach to this issue.

An important challenge in European river basin management is to develop strategies and solutions for coping with past and present sediment contamination. This can best be achieved by further reduction of contaminant emissions from point and diffuse sources. Further on, there will be a need to develop sediment management guidelines based on a multidisciplinary, coordinated and standardised approach. This might overcome the scattered responsibilities for sediment management and improve knowledge and experience of it in dredging administrations. Finally, efforts need to be made to increase public awareness of sediments as constituting an essential element of river basin ecosystems.

These are the reasons that the demand-driven European Sediment Research Network was set up. Sustainable sediment management at the river basin scale is an extremely challenging goal, which can only be achieved by establishing and maintaining a delicate balance between the interests of the environment, the society and the economy. SedNet has successfully shown possible ways forward to achieve this goal and has triggered a broad and intensive discussion in Europe on the role and importance of sediments in the environment [29, 30]. Whilst the EU-funded project ended in 2004, the network itself still exits and is active.

Besides these activities there are a few working groups in Europe and world-wide dealing with the specific issues of dredged material management. Most prominent among them are the Central Dredging Association (CEDA) and the International Navigation Association (INA, formerly PIANC). Both of them have issued a number of practical and helpful reports offering guidance to environmentally acceptable management of dredged material [31, 32].

2.10. Conclusions

- On a European level, management of aquatic sediments and dredged material is at present not covered by adequate regulations.

- In part this management is considered in different European directives, among others the Waste Directive. These regulations seem not to take adequately into account the specific sources and characteristics of aquatic sediments and dredged material ranging from clean, non-toxic sand to highly contaminated waste.
- Generic guidelines for the management of coastal dredged material have been developed by the international maritime conventions of London, OSPAR and Helsinki.
- On that generic basis, European member states have successfully developed their specific national guidelines for handling dredged material in coastal waters. This includes the derivation of action values helping the sediment managers to decide on the steps to be taken.
- A joint international approach for developing action values for dredged material does not yet exist. However, they are under discussion in the OSPAR Convention.
- To overcome the gaps in joint international management of sediments and dredged material, two projects, among others, have been initiated in Europe: A "Dutch–German Exchange on Dredged Material" (DGE) working group and a European sediment network (SedNet) for developing environmentally, socially and economically viable practices of sediment management on the river basin scale. There is also small number of European and world-wide working groups cooperating efficiently in issuing helpful guidance specifically on environmentally sound management of dredged material.
- A debate should be initiated on whether a specific European directive for the management of sediments is to be developed or whether, and how, it could have a more prominent position in existing legislation, such as the WFD. Contrary to the present situation, such a directive could take into account the natural origin of sediments, their characteristics and their role in a water body inadequately covered by existing regulations.

3. Monitoring programmes for aquatic sediments

3.1. European programmes

On the international level, the European Environment Agency (EEA) plays an important role in the field of environmental policy and monitoring. It collects all kinds of environmental data, including sediment quality data, from EU member states (and other countries) and publishes quality status reports.

The latest "Environmental Assessment Report" [33] presents a chapter on water. Although it includes a description of observed water pollution, the term "sediment" appears only once. From this report it may be concluded that sediment monitoring is not yet carried out on a European level. The development of standardised guidelines for monitoring aquatic sediments or the implementation of a coordinated programme seems to be relatively low on the EEA agenda.

This raises concern for several reasons: Many of the contaminants reported (heavy metals, pesticides) are mainly concentrated in the sediment phase, from which they can spread to the entire water body. Secondly, contaminated sediments may lead to the bioaccumulation of contamination in biota living in and on sediment and thus contamination is hazardous to these benthic organisms. Furthermore, since sediment is an integral part of any water body and is a useful quality indicator, it should deserve the same attention as the water phase or aquatic biota.

The situation may change in the near future since a WFD group on Analysis and Monitoring of Priority Substances (AMPS, subgroup on sediments) is working on technical guidance for monitoring of water, biota *and* sediments. In 2004 they had sent out a questionnaire on sediment monitoring to EU member states and prepared a draft overview of current practises pertaining to sediment [34]. These findings will serve as a basis for including sediments in future monitoring procedures. Final protocols have to be available before the end of 2006, when EU member states will be obliged to start monitoring of Priority Hazardous Substances in their waters.

3.2. International maritime programmes

Contaminants in European coastal sediments are monitored within the programmes of two international conventions (Table 4). These activities are organized by the OSPAR and HELCOM commissions, whose working groups are dealing with the various aspects from scientific basics to evaluating the results of monitoring programmes.

Table 4: Sediment monitoring programmes of international maritime conventions for the protection of the environment involving European states

Name of convention	Convention area	Sediment monitoring programme
LONDON	Globally marine areas	None
OSPAR	North Sea and NE Atlantic	Monitoring ongoing, reports available
HELSINKI	Baltic Sea	Monitoring initiated

Reports are available from the OSPAR Commission [35]. The HELSINKI Commission has carried out monitoring of radioactive substances in sediments for many years but has not selected sediments as a monitoring compartment in their programme COMBINE [36, Part D-2]. However, guidelines were developed [36, Part B-13] and the collection of data on hazardous substances in sediments was initiated [37].

3.3. International river commission programmes

In recent years, not only has the number of international commissions for rivers and lakes in Europe increased, but so has the scope of their monitoring programmes and the volume of published findings. Overviews of European transboundary cooperation in monitoring are available from the various commissions (see Annex 2) but also from independent organisations such as the European Rivernet [38] and the International Water Assessment Centre [39, 40]. The information available in the media shows that several commissions have established a monitoring programme for sediments, at present exclusively for the chemical analysis of contaminant concentrations.

Joint collection of sediment data in these groups is not necessarily based on standardised and strictly followed guidelines for sediment sampling, sample conservation and chemical analysis. A Dutch–German study of Rhine data [41] revealed that differences in these parameters render comparison of data from different commission members and a subsequent assessment of sediment quality difficult. Without improving the development and standardisation of guidelines for sediment monitoring, the assessment of sediment quality data and the subsequent risk assessment on a river-basin scale will remain problematic. On the other hand, reliable guidance for the institutions involved may increase the acceptance and utilisation of sediment monitoring.

Sediment characterisation on the basis of bioassays started several years ago and since then has reached a high level of scientific expertise. However, these assays have not been used in regular monitoring by any of the commissions. This step will become necessary in order to complement and improve the present monitoring of contaminant *concentrations* in sediments by determining contaminant *effects*. It will become all the more important as the European Water Framework Directive aims at *a good ecological status* of waters, which will also require a good ecological status of the sediments. (For an in-depth discussion on the relation between ecological status and bioassays see Chapter 6 (P. den Besten)).

3.4. National programmes

The previous chapter shows that a multitude of sediment monitoring programmes are in operation in Europe on the national and regional levels. Inventorying and, in particular, comparing these programmes can be problematic on two accounts: Obtaining a complete overview would be time-consuming, and the comparative assessment would be hindered by the great diversity of the programmes. For example, in all sixteen federal states in Germany, some kind of sediment monitoring is going on, although it is poorly coordinated and lacks standardisation. With no obligation to report sediment monitoring data to the European Environment Agency, sediment monitoring clearly has a low priority.

Before sediment monitoring is coordinated on a larger (river-basin) scale, it may prove helpful to collect the experience gained with national and regional programmes. Such an overview can serve as a basis for standardising scientific and administrative protocols from sampling and analysis to quality assessment and reporting formats.

3.5. Difficulties and advantages of sediment monitoring

The overview of sediment monitoring programmes in Europe reveals "that a wide range of different approaches is currently adopted towards sediment monitoring" [34]. Programmes are mostly lacking on the international level and are carried out - if at all - only on the national or regional scale.

Comparing international monitoring activities, it would seem that monitoring sediments is more difficult than monitoring water. The reasons may in part stem from the way sediments are formed and their various characteristics, which differ considerably from those of water:

- Recent sediment forms over a period of time when bed load material and suspended particulate matter settle in suitable areas.
- Disposal and re-suspension (erosion) of sediment occur reversibly depending on hydraulic conditions, in particular current velocity. These hydraulic conditions may vary within a short period of time and distance, making sediment layers much more inhomogeneous than water samples.
- The inhomogeneous character of sediment layers renders representative sampling as difficult as water sampling. However, sedimentation rate and grain size distribution are two of the important differentiating characteristics compared to water.

The reasons outlined above explain why monitoring sediments may require more sophisticated knowledge than monitoring water, and that interpretation of quality data obtained is far from trivial. Accordingly, sediment monitoring requires additional scientific knowledge, special sampling techniques, and harmonisation of monitoring and quality assessment, particularly if such monitoring stretches beyond regional borders.

On the other hand, the advantages of sediment monitoring may compensate for the drawbacks:

- Through their formation over a period of time, sediments may serve as a "long-term memory" of contaminant loads in a water body. Investigating sediments provides a retrospective history (trend) of contamination that cannot be obtained with water samples.
- A large number of dissolved hazardous substances (heavy metals, organic contaminants) have a strong tendency to be adsorbed onto the surface of solids. The suspended particulate matter always present in a water body acts as a scavenger for such substances. As a result, these substances may be difficult to detect in the water phase at a given time, possibly rendering an incomplete picture of the contamination situation.
- Since sediment formed from suspended particulate matter concentrates trace substances, their detection is easier and more sensitive.
- The relation of contaminants in sediments to those in biota living on or in the water bottom (benthos) is often more direct than with the water phase. Bioaccumulation and other adverse effects on benthos can only be studied and assessed by including sediments in the investigations. Therefore, an evaluation of the ecological state as foreseen in the Water Framework Directive must include data on sediment contamination.

3.6. Conclusions

An overview of sediment monitoring programmes in Europe shows that such programmes are mostly lacking on the international level and are carried out – if at all – only on the national or regional scale. The findings can be summarised as follows:

- Since no obligation exists for reporting sediment quality data to the European Environment Agency, sediment monitoring generally has had a low priority on the European level. This situation may change when EU member states, due to the Water Framework Directive, have to start status and trend monitoring of Priority Hazardous Substances in sediments.
- Only one international maritime convention (OSPAR) currently operates a regular monitoring programme, another one (HELCOM) is in preparation.
- A few of the international commissions for transboundary rivers are operating a sediment monitoring programme, among them the commissions for the Rhine, the Elbe and the Scheldt.
- On the national and regional scales, a considerable number of regular river sediment monitoring programmes are in operation. However, comparing

these data presents problems since very few of the monitoring programmes are standardised.

- Sediment monitoring seems to be more problematic than water monitoring. It therefore follows that standardisation of sediment monitoring beyond regional borders requires greater efforts than for water. This can perhaps be explained by how sediments are formed and by the differences between water and sedimentation monitoring.

- The relation of contaminants in sediments with biota living on or in the water bottom (benthos) is in many cases much more direct than in the water phase. Bioaccumulation and other adverse effects on benthos can only be studied and assessed by sediment investigations. Therefore, a statement on the ecological state as foreseen in the Water Framework Directive must include data on sediment contamination obtained by monitoring.

- The European Water Framework Directive (WFD) shows a new approach aiming at a river management on the basin scale. Since sediments are inseparably part of the intended "good ecological status" of a river, sediment monitoring within river basins is to be included in the development and implementation of the Water Framework Directive.

4. Recommendations

Specific guidelines for sediment management

Because of the natural origin, the characteristics and the role of sediments (including dredged material) in water bodies, it seems appropriate to think about their re-definition in the context of waste. For the same reason, developing specific guidelines for the management of dredged material has been proposed. A debate should be initiated on whether a specific European directive for the management of aquatic sediments should be developed. Contrary to the present situation, such a directive could take into account the specific characteristics of sediments inadequately addressed by existing EU regulations.

Polluter pays principle

It will be a major improvement for the entire aquatic environment when the "polluter pays" principle becomes a firm political objective in river basin-wide sediment management. Including it in the effective control of contaminants at the source in all concepts and regulations will be a great step forward in improving sediment quality and hence that of the entire aquatic environment.

European Water Framework Directive (WFD) and sediments

The WFD proposes a new approach aiming at a river management on the basin scale. Since sediments are inseparably part of the intended good ecological status of a water body, sediment monitoring within river basins is to be included in the development and implementation of the Water Framework Directive.

In many cases the relation of contaminants in sediments with biota living on or in the water bottom (benthos) is a much more direct one and closer to the actual situation than measurements of the water phase. Bioaccumulation and other adverse effects on benthos can only be studied and assessed by including sediment investigations. Consequently, a statement on the ecological status as requested in the Water Framework Directive must include data on sediment contamination obtained by monitoring.

Sediment monitoring and quality assessment

On the national or regional scale, comprehensive monitoring of aquatic sediments is underway. However, comparing these data is problematic since very few of the monitoring programmes are internationally harmonised. Apparently, monitoring sediments presents its own requirements due to the way sediments are formed and to a number of their characteristics, which are quite different from the characteristics of water.

Consequently it follows that standardisation of sampling and analysis beyond regional borders is a prerequisite for successful sediment monitoring. The same holds true for the development of standardised action values as a joint basis for the subsequent quality assessment.

Annex 1

Occurrence of the keyword „SEDIMENT" in the EU Water Framework Directive

Source: DIRECTIVE OF THE EUROPEAN PARLIAMENT AND OF THE COUNCIL 2000/60/EC ESTABLISHING A FRAMEWORK FOR COMMUNITY ACTION IN THE FIELD OF WATER POLICY

Luxembourg, 23 October 2000, Document 1997/0067(COD), C5-0347/2000, LEX 224, PE-CONS 3639/1/00, REV 1, ENV 221, CODEC 513, **152 pages**

Page	Text
21	35) "Environmental quality standard" means the concentration of a particular pollutant or group of pollutants in water, *SEDIMENT* or biota which should not be exceeded in order to protect human health and the environment.
53	7. The Commission shall submit proposals for quality standards applicable to the concentrations of the priority substances in surface water, *SEDIMENTS* or biota.
98	Biological quality elements
table 1.2.1 Definitions for high, good and moderate ecological status in rivers	Phytoplankton - Good status:
	There are slight changes in the composition and abundance of planktonic taxa compared to the type specific communities. Such changes do not indicate any
	accelerated growth of algae resulting in undesirable disturbances to the balance of organisms present in the water body or to the physico-chemical quality of the water or *SEDIMENT* .
98	Biological quality elements:
table 1.2.1: Definitions for high, good and moderate ecological status in rivers	Macrophytes and phytobenthos - Good status
	There are slight changes in the composition and abundance of macrophytic and phytobenthic taxa compared to the type-specific communities. Such changes do not indicate any accelerated growth of phytobenthos or higher forms of plant life resulting in undesirable disturbances to the balance of organisms present in the water body or to the physico-chemical quality of the water or *SEDIMENT*.

99	Hydromorphological quality elements:
table 1.2.1: Definitions for high, good and moderate ecological status in rivers	River continuity - High status
	The continuity of the river is not disturbed by anthropogenic activities and allows undisturbed migration of aquatic organisms and *SEDIMENT* transport.
101	Biological quality elements:
table 1.2.2: Definitions for high, good and moderate ecological status in lakes	Phytoplankton - Good status
	There are slight changes in the composition and abundance of planktonic taxa compared to the type-specific communities. Such changes do not indicate any accelerated growth of algae resulting in undesirable disturbance to the balance of organisms present in the water body or to the physico-chemical quality of the water or *SEDIMENT*.
101	Biological quality elements
table 1.2.2 Definitions for high, good and moderate ecological status in lakes	Phytoplankton - Moderate status
	Biomass is moderately disturbed and may be such as to produce a significant undesirable disturbance in the condition of other biological quality elements and the physico-chemical quality of the water or *SEDIMENT*.
110	1.2.6 Procedure for the setting of chemical quality standards by Member States
	In deriving environmental quality standards for pollutants listed in points 1 - 9 of Annex VIII for the protection of aquatic biota, Member States shall act in accordance with the following provisions. Standards may be set for water, *SEDIMENT* or biota.

Annex 2

Table: Some major international commissions in Europe for the protection of transboundary rivers and lakes

River basin	Name	Internet address
Danube	International Commission for the Protection of the Danube River	www.icpdr.org
Elbe	International Commission for the Protection of the Elbe	www.ikse-mkol.de
Ems	Ems Commission	
Lake Geneva	International Commission for Protection of Lake Geneva	www.cipel.org
Lake Constance	International Water Protection Conference for Lake Constance	www.igkb.org
Meuse	Commission Internationale pour la Protection de la Meuse	www.cipm-icbm.be
Moselle	Commission for the River Moselle	moselkommission.org
Moselle-Saar	International Commissions for the Protection of the Rivers Moselle and Saar	www.iksms-cipms.org
Oder	International Commission for the Protection of the Oder	www.mkoo.pl
Oder	German-Polish Border River Commission	
Rhine	International Commission for the Protection of the Rhine	www.iksr.org
Scheldt	Commission Internationale pour la Protection de L'Escaute (Schelde)	www.icbs-cipe.com
UNECE	United Nations Economic Commission for Europe	www.unece.org/ env/water
	Border River Commission between Finland and Sweden	www.environment.fi
	Border River Commission between Finland and Russia	www.ymparisto.fi
	Finnish-Norwegian Commission on Border Water Systems	www. environment.fi

Note: More information on cooperation on transboundary waters is found under www.rivernet.org and www.iwac-riza.org

References

1. ISO (1996): ISO 11074-1, Soil quality - Vocabulary Terms and definitions relating to the protection and pollution of the soil
2. PIANC (2002): Environmental Guidelines for Aquatic, Nearshore and Upland Confined Disposal Facilities for Contaminated Dredged Material. PIANC International Navigation Association, Working Group EnviCom 5, Report
3. SedNet (2006): for information see www.SedNet.org
4. WFD AMPS (2004-a): Discussion Document on Sediment Monitoring Guidance for the EU Water Framework Directive. Version 2, 25 May 2004, Working Group on Analysis and Monitoring of Priority Substances (AMPS), Subgroup on Sediment
5. OSPAR (2004): Revised OSPAR Guidelines for the Management of Dredged Material. Doc. BDC 04/5/13
6. LC (2000): Specific guidelines for assessment of dredged material. Doc. LC 22/5/Add.1
7. PIANC (2004): Environmental Risk Assessment in Dredging and Dredged Material Management. PIANC International Navigation Association, EnviCom Working Group 10, Report
8. ISO (1999): ISO 11074-3, Soil quality - Vocabulary Terms and definitions relating to soil and site assessment
9. UNECE (2004): What Is It, Why It Matters? United Nations Economic Commission for Europe, Water Convention Secretariat, Geneva
10. Köthe H, de Boer P (2003): Dredged Material and Legislation. Dutch–German Working Group on Dredged Material, Part 1, Federal Institute of Hydrology, Koblenz, Germany
11. EC (1975): European Framework Directive on Waste, Council Directive of 15 July 1975 on waste (75/442/EEC); modified by Decisions 91/156/EEC and 96/350/EC
12. EC (2000-a): European Waste Catalogue (2000/532/EC), Council Decision of 3 May 2000
13. EC (1999): European Landfill Directive (1999/31/EC), Council Decision of 26 April 1999
14. EC (2000-b): Directive of the European Parliament and of the Council (2000/60/EC) establishing a Framework for Community Action in the Field of Water Policy. 23 October 2000
15. EC (2002-a): Commission Communication (2002): "Towards a Thematic Strategy for Soil Protection", document (COM(2002) 179)
16. ESB (2005): European Soil Bureau: Future perspectives (http://eusoils.jrc.it/esbn/esbn_future.html)
17. EUSOILS (2006): Land Management and Natural Hazards Unit: http://eusoils.jrc.it/index.html
18. Forum Europa (2004): Working group final reports, http://forum.europa.eu.int/Public/irc/env/soil/library?l=/
19. Defra (2006): EU Thematic Strategy for Soil Protection, www.defra.gov.uk/environment/land/soil/europe/
20. Commission of the European Communities: Proposal for a Directive of the European Parliament and of the Council establishing a Framework for Community Action in the field of Marine Environmental Policy (Marine Strategy Directive). Brussels, 24.10.2005, COM(2005) 505 final
21. Commission of the European Communities: Communication from the Commission to the Council and the European Parliament: Thematic Strategy on the Protection and Conservation of the Marine Environment and Proposal for a Directive of the European Parliament and of the Council establishing a Framework for Community Action in the field

of Marine Environmental Policy (Marine Strategy Directive), Impact Assessment. Brussels, 24.10.2005, COM(2005) 504 final

22. HELCOM (1994): Guidelines for the Disposal of Dredged Spoils. HELCOM Recommendation 13/1, adopted 1992 (under revision)
23. OSPAR (2003-a): Contracting Parties' National Action Levels for Dredged Material. Working Group Environmental Impact of Human Activities (EIHA), doc. EIHA 03/2/3-E
24. GIPME (2000): Guidance on Assessment of Sediment Quality. Working Group Global Investigation of Pollution in the Marine Environment, International Maritime Organization, London, Publ. No. 439/00
25. ICPR (1996): Nationale Anforderungen an die Aufbringung und Ablagerung von Baggergut [National requirements for the disposal and relocation of dredged material]. International Commission for the Protection of the Rhine, Doc. PLEN 8/96, rev. 05.07.96 (1996)
26. Bergmann H and V. Maass (2004): European sediment regulations: Gaps and bridges. Presentation, Final SedNet Conference, Venice, Italy, 24-25 November 2004
27. EC (2004): Directive 2004/35/EC of the European Parliament and of the Council of 21 April 2004 on environmental liability with regard to the prevention and remedy of environmental damage. J. European Union, 30.4.2004, L 143/56
28. DGE (2006): for information see www.htg-baggergut.de/Downloads/
29. Salomons W. and J. Brils (Eds.) (2004): Contaminated Sediments in European River Basins. European Sediment Research Network (SedNet), EC contract No.: EVK1-CT-2001-20002, 46 p., see www.SedNet.org
30. TNO (Ed.) (2004): The SedNet Strategy Paper - The opinion of SedNet on environmentally, socially and economically viable sediment management. European Sediment Research Network (SedNet), EC contract No.: EVK1-CT-2001-20002, 13 p., see www.SedNet.org
31. CEDA (2006): Central Dredging Association, for information see www.iadc-dredging.com
32. PIANC (2006): PIANC International Navigation Association, for information see wwww.pianc-aipcn.org
33. EEA (2003): Europe's Environment: The Third Assessment. Environmental Assessment Report No. 10, European Environmental Agency, Copenhagen
34. WFD AMPS (2004-b): Summary of responses to questionnaire on sediment monitoring. Draft, 23 March 2004, Working Group on Analysis and Monitoring of Priority Substances (AMPS), Subgroup on Sediment
35. OSPAR (2000): OSPAR Commission 2000. Quality Status Report 2000. OSPAR Commission, London. 108 + vii pp. ISBN 0 946956 52 9
36. HELCOM (2003): Manual for Marine Monitoring in the COMBINE Programme of HELCOM
37. ICES (2006): www.ices.dk/datacentre/accessions/, HELCOM data
38. ERN (2004): European Rivers Network, Internet publication (www.rivernet.org)
39. IWAC (2004-a): International Water Assessment Centre, Monitoring Tailor-Made IV, 15–18 September 2003, St. Michielsgestel, The Netherlands (www.iwac-riza.org)
40. IWAC (2004-b): Assessment Practices and Environmental Status of Ten Transboundary Rivers in Europe, www.iwac-riza.org
41. Bergmann H, ten Hulscher D, Eisma M (2005): Hazardous Substances in Dredged Material. Working Group Dutch-German Exchange on Dredged Material (DGE), Part 3, AKWA/RIZA, Lelystad, The Netherlands (www.htg-baggergut.de/Downloads/)

Sustainable Management of Sediment Resources: Sediment Risk Management and Communication 233
Edited by Susanne Heise

Risk Perception and Risk Communication

Gerald Jan Ellen[a], Lasse Gerrits[b] and Adriaan F. L. Slob[a]

[a] *TNO Environment and Geosciences, Van Mourikbroekmanweg 6, Postbus 49, 2600 AA Delft, The Netherlands*
[b] *Erasmus University Rotterdam, Department of Public Administration, Postbus 1738, 3000 DR Rotterdam, The Netherlands*

1. Introduction

The notions 'stakeholders', 'perception', and 'communication' have appeared several times so far in this book. The ideas and motives for stakeholder involvement can be found in Chapter 2, *'Sediment Management Objectives and Risk Indicators'* and an in-depth discussion can be found in "Sediment Management at the River Basin Scale" by Philip N. Owens (Ed.), this book series. This chapter is dedicated to two themes: risk perception and risk communication. The societal dimension of risk-management of sediments has hardly been discussed in scientific literature so far. This is a first step towards a more comprehensive overview of this issue. Stakeholder involvement is mandatory because of recent legislation concerning rivers [1], most notably the Water Framework Directive from the European Commission. With stakeholder involvement comes the need to understand the perception of risk, as opposed to scientifically assessed risk, and recognition of the benefits of communication with stakeholders.

Risk management of sediments can be viewed on a variety of scales, which can be site-specific or basin scale, as has been pointed out earlier in this book. In order to achieve sustainable development and thus attain a sound allocation of resources, risks at specific sites need to be assessed with respect to their interaction with and contribution to other sites upstream and downstream - all the way up to the entire river basin. Whatever the scale, sediment management options have an effect on specific sites. In other words, the measures taken will become concrete and visible at specific places and this might provoke public resistance. For example, when the river Elbe in Germany was deepened in order to allow larger ships to enter the port of Hamburg, this appeared to have an impact on the safety of dams along the Elbe river and on marinas along the Elbe filling with sediments [2]. In other words, sediment management options cause impacts on sites, regardless of the scale, which means that **lay people**, people who have no or very little (expert)knowledge of the subject, are affected and consequently become stakeholders. Following the principle of stakeholder involvement, these people need to be involved in some way and to some extend in the process of decision making on sediment management options. Risk assessment is part of the process of finding and implementing solutions for sediment issues.

Risk assessment is done because adverse effects for social and physical systems are expected. Risk assessment has long been perceived as a mere technical procedure, a complicated one, but in the end within the domain of scientific determination. However, over time it has become clear that a gap exists between this so-called scientifically assessed risk and what people perceive as dangerous or risky [3,4].

Contrary to technical risk calculated by experts, perceived risk can raise people's resistance against sediment management options. The case of the Ingense Waarden, as shown in box 3 in Chapter 2 of this book, is one of the many examples of how lay people can raise strong opposition against sediment management – even to the point where a decision will be delayed or abandoned. Lay people can become very concerned about matters they perceive as risky, and this includes contaminated sediments [5]. At the same time, the general awareness of what sediment is, what contaminated dredged material is, and the potential dangers this poses is very low. More attention is usually paid to the risks caused by sediments when an incident occurs. Such an incident might be the detection of high levels of contamination in dredged material that has been used to level up polders or as building material. But it can also concern the sudden announcement of the (local) government to build a dredged material disposal site near a populated area [5]. The point here is that the perception of risk on the part of non-experts does not develop in synchrony with the risk assessment of experts. At the same time, both groups – experts and lay people – will have to deal with each other: the sediment management option affects lay people, and their attitude affects the progress of the deriving of the management solution. The experts, therefore, have to take risk perceptions into account.

However, both experts and politicians hesitate to engage the process of stakeholder involvement. From the experts' point of view, scientific risk assessment might be superior to perceived risk, the latter often considered as irrational. And politicians might fear that management decisions could be openly rejected, delaying the process, increasing costs and undermining their authority. Moreover, both experts and politicians might be convinced that lay people lack the ability to participate in environmental decision making. This in turn leads to distrust among the lay people with regard to the motives of the experts and the politicians and their reasoning.

Understanding how risks are perceived and what governs the public assessment is necessary to develop a concept on how to deal with and convey information on different management alternatives, and how and to what extent lay people could be included in decision making. The first sections in this chapter (sections 2-4) discuss risks in society and the differences between scientifically assessed risk and perceived risk. The place of risks in society has changed over the past years in that people in developed countries no longer need to worry about their very basic needs. Risk assessment has always encompassed a scientific approach, but with the growing concern people have of avoiding risks, the concept has become multi-facetted and the notion of perceived or subjective risk has come to the forefront. The roots of perceived risks and the subsequent plurality are explained. Section 5 is dedicated to communication with stakeholders, bearing in mind the different perceptions on risk. Finally section 6 contains the conclusions of the chapter.

2. Society's growing concern with risk

The thesis of the German sociologist Ulrich Beck is that contemporary societies are characterised by risk having taken a central position, hence the concept of the so-called 'risk-society'. In this society, shaping risks systematically follows shaping values. Since people have provided themselves with a sound fundament for sustaining basic life needs, for example through a steady supply of food and social benefits, the focus has changed from running risks to provide this fundament to avoiding risks in order to defend it [6]. Because modern industrialised societies have drastically limited the occurrence of external hazards, risks have become dependent on decisions; they are industrially produced.

This in turn means that risks have become more intangible. They are now subject to the debate on whether running the risk is necessary given the possible advantages of doing so. This debate has become more and more open to stakeholders who were previously not involved. For example, the disastrous flooding that hit the south west of the Netherlands in 1953 led to the construction of a number of dams and dykes in the Zeeland delta. More recently, experts have found it desirable to open reclaimed land for flood control areas and the (re-)development of nature. In the past, experts could use technical data to show and convince other stakeholders of the necessity of a plan. Nowadays, stakeholders no longer attach much value to this because they are more afraid of the image of the 1953 flood than of the technical analysis showing that the risks are negligible. Instead of running a possible risk in order to obtain something (flood control and ecological development), people choose to avoid risk in order to safeguard what they have (dykes and farmland). In other words, the perception of risk has become stronger than the assessed risk, the discussion of risks has entered the public debate, and the outcome depends more and more on the attitude of larger groups of stakeholders.

3. Objectivism and constructivism

It can be noted from the example above that risks have two dimensions. On the one hand, *risks are scientifically assessed*. An extensive overview of this can be found in Chapter 5 of this book ("*Risk Assessment Approaches in European Countries*"). On the other hand and in addition to this type of risk assessment, where *scientific objectivism* is promoted, is the notion of *perceived risk*, or *subjective risk*. How do these two relate to each other?

Risk analysis consists of three interconnected components: risk assessment, risk management and risk communication. Risk analysis has become an established field of research since the early 1970s, where it was originally rooted in engineering and decision sciences. The notion of 'risk' was mainly confined to the natural sciences, in which probabilistic risk assessment as a method dominated. Since the 1980s, 'risk' has gained importance as a concept in assessment activities, especially in technology assessment. Risk is still central to societal debates on technology. In the last decade, the concept of risk has also been increasingly used to refer to environmental phenomena, such as ozone depletion and climate change. Risk analysis is furthermore becoming part of assessment activities for the purpose of environmental policy making. Since the beginning of the 1990s, using the concept of risk to address general aspects of decision making in modern society has been advocated. These recent developments seem to have launched a trend in risk analysis of extending beyond a specific case. Contemporary risk analysis can be described as a scientific approach towards risk, used for public policy making on technological, environmental and health issues [7] and is an active field involving many different disciplines, e.g. mathematics, statistics, systems analysis, psychology, political sciences and even philosophy.

Within the evolving field of risk analysis, the importance of and approach towards risk perception have also changed. The concept of risk perception has evolved from being considered a one-dimensional series of measurements into a multidimensional concept that involves people's beliefs, attitudes, judgments and feelings, as well as the wider social or cultural values and dispositions adopted towards hazards and their benefits [6,8].

The theoretical development can be divided into two main approaches. The first approach focuses primarily on the quantification of risk acceptance through revealed preferences. This originated with the work of Starr (1969) in which social benefits were correlated with the number of accidental deaths arising from the application of technological developments [9]. The main proposition of this approach is that objective and subjective risk can be maintained as separate, i.e. risk acceptance can be objectively assessed and measured from revealed preferences by using historical data on fatalities in the public use of technology. Second, this approach suggests that risk can be assessed objectively through statistical methods that calculate the probability of loss. Authors advocating this theory are usually described as objectivists [8] and the approach as the objectivist approach.

Empirical research by Van Asselt has shown that there is a gap between scientific estimates and the estimates by lay people. Starr introduced the distinction between objective and perceived risk to discriminate between the scientific definition and the perception of lay people. Objectivists consider this gap between the experts' view and the view of lay people as 'simple misperceptions, biases or plain deviousness' [7]. The objectivists consider perceived risks as inherently wrong, because lay people often overestimate involuntary risks [7]. The 'gap' between lay people and experts can only be bridged by putting enough effort into communication and involvement, and by putting the same emphasis on lay perception as on technical knowledge and data used to estimate the risks.

Why should one bother to do so? In a world where lay people have learnt to become more involved, and where power has become increasingly shared with them, the main motive for listening to their opinions and views is that the chance that they will use their obstructive power increases if they are ignored [10]. Opposition to plans will grow, as citizens do not feel that they are taken seriously. This opposition can be persistent and have serious consequences for the progress of the decision making process, which decision makers will want to avoid at all costs. So the conclusion can only be that citizens should be taken seriously, and hence that their views must be incorporated into the risk assessment.

Objectivists also have their own view on risk analysis. Risk assessment is considered as the 'solid' analysis that measures the 'real' risks. Risk management comprises the administrative and political procedures to deal with risks. It is the process of deciding what to do about risks. Objectivists assume that the right expert estimations would be enough to settle questions about the acceptability of risks. Risk communication from this perspective is considered to be used instrumentally: communicating the experts' view to inform and reassure the public. This means for sediments that only information about sediment management will be sent to the stakeholders, for instance, the fact that the dredged material is not contaminated or that the deepening of a river will not have an impact on the ecology of the river. This is usually done through websites, brochures and advertisements. Later in this chapter, it will be argued that there are other approaches as well.

The **constructivist** school of thought is a response to the objectivist school. It questions the objectivists' fundamental premises. According to constructivists, there is no objective definition of risk. Risks are inherently subjective. Consequently, there are various definitions that are more or less appropriate, depending on the problem and the context. Risks are socially constructed, so there is not one best way to draw a distinction between objective and perceived risks [7].

Douglas (1970) first introduced the idea that the definition and perception of risks are rooted in social and cultural conditions [11]. Multiple legitimate risk definitions exist due to a wide variety of psychological factors and social and cultural circumstances. Constructivists say that risk analysis never should be an exercise performed by scientists alone, because every individual produces his or her own selected view. According to constructivists, societal stakeholders should participate in risk analysis. Participatory processes should be an integral part of risk management [7].

The main characteristics of the objectivist and the constructivist schools of thought in risk analysis are summarised in Table 1 [7].

Table 1 Characteristics of the two major schools of though in risk analysis according to Van Asselt [7]

Objectivists	*Constructivists*
Science is value-free (i.e. positivism).	Science is entirely social (i.e. social or cultural relativism).
Distinction between objective and perceived risks.	Risk is a social construct. There is no objective definition of risk
Objective risks are measurable in terms of probability and utility.	Risk analysis should involve qualitative factors that are difficult to measure.
Risk assessment and risk management have to be separated.	Risk assessment and management are inseparable activities with value differences are at the core.
Accurate expert estimations can settle risk issues.	Participatory processes are needed to manage risk issues.

The objectivist view dominated risk analysis until the 1990s. Since then it has lost ground to the constructivist approach. This does not mean that objectivistic approaches have disappeared. Objectivistic means of risk assessment are still widely used in Europe, as is shown in Chapter 5, and it has not become obsolete. But the constructivist approach has had a major influence and the majority of risk analysts seem to agree on the following premises [7]:

- Experts perceive risk in a different way than do lay people.
- Risk analysis is not a purely objective process: facts and immaterial values merge frequently.
- Cultural factors affect the way people assess and perceive risk.

This change of approach is also taking place in the realm of environmental management. Therefore, this approach is examined more closely here. The following section is dedicated to alternative approaches that fit into the constructivist school.

4. Plurality and risk

There are differences between the individual perspective and the collective perspective. They are not necessarily the same. Risk management of sediments at the river basin scale, or even at specific sites, exceeds the individual's interests but at the same time, the collective interests are the aggregate of those individual interests. Therefore, attention should be paid to the collective perspective on risks. As to what determines this perspective, one should be aware of the importance of culture in shaping individual stances and consequently collective perspectives.

The cultural theory applies in these instances. Culture – with its written and unwritten rules of conduct, its 'invisible' coordination in society and influence on individuals' behaviour – is seen as the most influential factor for the orientation of people towards risk [10]. At the basis of this theory is Mary Douglas's observation that in different cultures, two basic dimensions of social organisation are present: a group dimension and a grid dimension. The group dimension describes the extent to which individual behaviour is influenced by group membership. A strong group membership leaves less room for individual behaviour, while weak group membership translates to strong individual behaviour. The grid dimension describes the extent to which individuals' or groups' behaviour is prescribed by rules.

The theory was developed by Thompson, Ellis and Wildavsky (1990) for policy analysis and distinguishes four basic types of social organisation by combining the group and grid dimensions [12]. These four types of social organisation, often described as 'rationalities' are named hierarchists, egalitarians, fatalists, and individualists (Figure 1) [13].

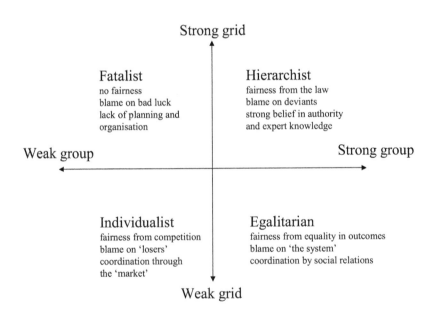

Figure 1. A grid-group map by Devilee [13]. (See text for explanation of 'group' and 'grid' and further description of each of the 4 different categories)

These four different categories can be described as follows:

Hierarchists: strong group, strong grid. Hierarchists have a strong group relation and strong binding prescriptions. The collective level is very important for hierarchists. They believe in strong regulation, have a high esteem for authority and judge expert knowledge as very high. Fairness has to be derived from the law. Blame is put on deviants who do not live up to the rules and procedures.

Egalitarians: strong group, weak grid. Egalitarians have a strong group relation with minimal prescriptions. The group is maintained by strong interpersonal relations. Equality is a strong value for this group. Egalitarians do not like authority. Fairness is derived from equality in result. Blame is put outside the group, most of the time on 'the system'.

Fatalists: weak group, strong grid. Fatalists are subject to strong binding prescription and have no group binding. The life of the fatalist is more or less organised from the outside: they have very little influence on it themselves. Fairness does not exist in their view and blame is put on bad luck. Fatalists cannot plan nor organize things in a way that is good for them.

Individualists: weak group, weak grid. Individualists live in a social context where there is neither group incorporation nor strong prescriptions from the group. Coordination in society comes from 'the market', where individuals negotiate with each other and enter into transactions. Fairness consists of equality of opportunity. As individualists are living in a very competitive environment, blame is on those who cannot compete, so on 'personal failure'.

The theory has been extended to the 'myths of nature' by Schwarz & Thompson [14], which transfers the grid-group typology into different ways how people look at the environment and physical systems. Different orientations of the perspectives towards risk can be deduced based on this (see Figure 2).

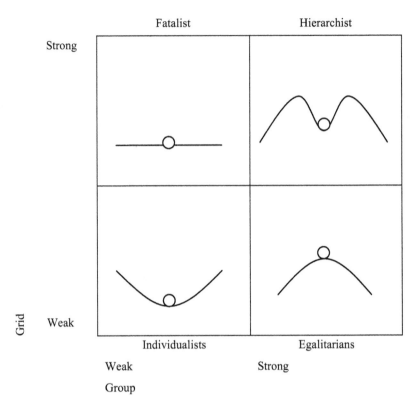

Figure 2 The cultural theory and risk (after Thompson et al., 1990): *fatalists* feel that risks are unmanageable, *hierarchists* feel that risks are acceptable within boundaries, *egalitarians* feel that risks have to be avoided at all costs and finally *individualists* feel that there is no reason to avoid risks.

The different perspectives on sediment of the stakeholder groups were analysed in a case study in the Netherlands. In-depth-interviews with representatives of the stakeholders to determine their perspectives on risk lead to the finding of three different perspectives: clearly illustrating the active cultural perspectives:
User (relates to the individualists perspective). This perspective is characterised by a short-term vision on sediment. From the Users' perspective, sediment can be seen as a useful resource, for example as a fertilizer, construction material or a resource for elevating land. But sediment can also be an obstacle, which needs to be removed to make waterways more accessible for recreational and/or commercial shipping. From this point of view, there are no risks, only challenges that can be resolved by applying technology. The quality of the sediments is only important if it contributes to the usefulness of the sediment, or if it creates a problem in handling the sediments (for example if these are contaminated). The Users' approach to policy and regulation is of a practical

nature; every situation needs to be viewed within its own context, and the 'rules' should be applied accordingly.

Controller (relates to the hierarchists perspective). The most important aspect of sediment from the controllers' perspective is to avoid societal risks. Sediment can pose a threat to society (mainly) in two ways: flooding and contamination. From this perspective, it is therefore very important to manage both the risks of flood and the risks of contamination for environmental and public health. The timeframe is set for the short-to-medium term. So the controllers anticipate future developments, but also keep a close eye on the current situation (focus on risk management). Information gathering and research are essential to this perspective. Research is used to deal with uncertainty of any kind. There is never enough information on an issue. The same goes for policy and regulation, which is seen as a framework for retaining control over a (seemingly controllable and rationalised) situation.

Guardian (relates to the egalitarian perspective). From this perspective, long-term timeframes are important. Guardians see sediment as part of the ecosystem and it should be handled with care. The attitude towards risk, therefore, is one of avoiding risks at all costs. The reason for this is that changing the present situation can have (unintended) consequences for the future. The focus is mainly on the quality of sediments. Safety is also important, but only when no alternative is available can changing the ecosystem be allowed. If there is a situation in which this is unavoidable, than it should be done with utmost caution. The Guardians' goal is to obtain 'natural' sediment, because this is an important aspect of the ecosystem, which influences water quality and the variety of flora and fauna. Policy and regulation should be directed towards making this goal attainable and preventing the intervention of short-term (economic) goals.

The fatalist perspective was not recognized in this case study. This is because this perspective is inherently passive in nature.

From these different perspectives, different perceptions of risks arise and this in turn leads to different approaches to the management of risks. Consequently, while objectivist risk assessment starts from the point of view that risks can be estimated and managed, the constructivist school of thought shows how different cultural perspectives influence the assessment of risks and the consequences of these assessments. In practice, people will view risks from different perspectives in different situations. But these perspectives make it clear that there are different attitudes towards risks related to sediment management and that communication concerning risks should address these differences because people will respond in different ways to risks.

5. Risk communication

Slob et al. [15] stresses that it is essential to communicate to the different perspectives in their own language or vocabulary, and to be aware of the 'blind spots' that are inherent to the different perspectives. 'Language' refers to certain key words that are often used by a particular perspective, which address subjects that are of great value to that perspective and that have unambiguous meanings to them. Blind spots, on the other hand, are words that are rarely used, are of little or no importance and have ambiguous meaning to a given perspective.

For a decision maker, it is not only important to know the different perspectives on sediments. Communicating to the different perspectives requires awareness of the different vocabularies and blind spots within those perspectives. As the message should be understood by all perspectives, communication should use language and words that have meaning for the perspectives involved and, therefore, should contain different variations of the message that the different perspective groups can understand (plural communication). Table 2 provides a number of examples of terms and blind spots for the different perspectives are given[*].

Table 2. Examples of language and blind spots of the different perspectives

	Language	Blind spots
User (Individualists perspective)	Challenge and profit	Long-term impact
	Technology	Ecosystem
	Pragmatic	Risk
	Costs	Control & regulation
Controller (Hierarchists perspective)	Government: control & regulation	Unusual, 'risky' solutions
	Danger/safety of sediments	Ownership of solutions
	Research (predict outcomes)	Costs are no 'hurdle'
Guardian (Egalitarians perspective)	Damage to nature/ecosystem	Economically viable
	Waste	Efficient solutions
	Risk	Short-term impact
	Regulation	Costs

---

[*] Some of the languages and blind spots described are the outcome of the SedNet workshop on Risk perception and Risk communication held in Athens on the 27th and 28th of November 2003.

These terms and blind spots can be used in communication with the different perspectives. Strong reactions might be expected when using 'their' language, and when giving information to or communicating with the different perspectives it is important to keep this in mind. The table also indicates the important observation that some of a perspective's language actually is the blind spot of the other. For instance, guardians have a blind spot for economic reasoning (economically viable, efficient), while this is the language of the user. This stresses the importance of using different terms that have meaning for all the perspectives when communicating with stakeholders.

The tools that can be used by decision makers also depend on the different levels that they would like the stakeholders to be involved in during the decision process. The necessity to actually involve stakeholders is not discussed here, because it has already been discussed in Chapter 2. The following overview (table 3) shows communication tools that can be used, based on the seven levels over stakeholder involvement by Pröpper and Steenbeek [16] (see also table 7 in Chapter 2).

Table 3. Means of communication

Governance styles within the scale of participation [16]	Means of communication
1. Closed authoritarian	None
2. Open authoritarian	Brochures, newsletters, internet sites, speeches, commercials
3. Consulting style	Group sessions, study groups, public gatherings, interviews, internet discussion
4. Participative style	Debating meetings, house calls, internet discussions, public gatherings with discussion.
5. Delegating style	Debating meetings, house calls, internet discussions, public gatherings with discussion.
6. Co-operative style	Create a common ground for discussion, for example by joint fact-finding. Mediation (depending on the situation).
7. Facilitating style	Create a common ground for discussion, for example by joint fact-finding. Mediation (depending on the situation).

Means of communication from the upper rows in the table can be used in lower rows but not the other way round. For example: means of communication applied in the co-operative style can be used in the open authoritarian style but the means of communication used in the latter cannot be used on the co-operative style as it would take away the co-operative character.

6. Conclusions

Risk perception and risk communication are discussed in this chapter. It is demonstrated why risk perception is important for risk communication. Risk assessment with respect to sediment management cannot be a technical procedure only. Instead, it should be acknowledged that there are different perspectives on risks, as shown in the constructivist school of thought. There are four extreme perspectives on risks: hierarchist, egalitarian, individualist and fatalist. In a research project based on a case study, three relevant perspectives on sediment management were observed: controller, guardian and user perspectives on sediment. Risk communication on sediment management should take into account the different languages and blind spots that occur within each perspective.

In summary:

- In sediment management issues, it is essential to respect the risk perception of all stakeholders, even when this does not comply with the scientifically estimated risk.

- Different stakeholders have different perspectives, meaning that they also have different views on risks. The different perspectives also have different vocabularies and blind spots, which should always be addressed in communication with these groups. Plurality in communication, which means communicating using the language of the different perspectives, is therefore very important.

- A diversity of communication tools that use different approaches, images and media that respect the language and blind spots of these different world views will help to reach the different stakeholders and to integrate them into the decision-making process, increasing support for decisions taken.

References

1. Ast JAv, Boot SP (2003): Participation in European water policy. Physics and chemistry of the earth 28(12-13): 555-562
2. (2004). Elbvertiefung mit Moderator, *Die Tageszeitung*: Hamburg
3. Renn O (2004): Perception of risks. Toxicology Letters 149: 405-413
4. Slovic P (2000): The perception of risk. Earthscan Publications Ltd.: London.
5. National Research Council (1997): Contaminated sediments in ports and waterways: Cleanup strategies and technologies. NRC Committee on contaminated sediments, marine board.
6. Beck U (1992): Risk Society. Towards a New Modernity. Sage Publications: London.
7. Asselt MBAV (2000): Perspectives on certainty and risk. The PRIMA approach to decision support. Kluwer Academic Publishers: Dordrecht.
8. Montalvo CC (2002): Environmental Policy and Technological Innovation: Why do firms adopt or reject new technologies? Edward Elgar Publishing: Cheltenham UK, Northhampton, MA.
9. Starr C (1969): Social benefit versus technological risk. Science 165: 1232-1238
10. Gerrits L, Edelenbos J (2004): Management of sediments through stakeholder involvement; The risks and value of engaging stakeholders when looking for solutions for sediment related problems. Journal of Soil and Sediments 4(4): 239-246
11. Douglas M (1970): Natural Symbols: explorations in cosmology. Barrie and Rockliff: London.
12. Thompson M, Ellis R, Wildawsky A (1990): Cultural Theory. Westview Press: Boulder.
13. Devilee JLA (2002): Decision making about waste facilities. An analysis of reactions of local residents in a risk society. Printpartners Ipskam: Amsterdam.
14. Schwarz M, Thompson M (1990): Divided we stand. Redefining politics, technology and social choice: Pennsylvania.
15. Slob A, Tielens T, As DMv, Ellen GJ (2003): Building trust together; study of consumer trust and food safety. TNO Built Environment and Geosciences.
16. Pröpper IMAM, Steenbeek D (2001): De aanpak van interactief beleid: elke situatie is anders. Coutinho: Bussum.

Sustainable Management of Sediment Resources: Sediment Risk Management and Communication   249
Edited by Susanne Heise

The Way Forward for Sediment Risk Management and Communication – A Summary

Susanne Heise[a], Sabine Apitz[b,c], Marc Babut[d], Helge Bergmann[e], Piet den Besten[f], Gerald Jan Ellen[g], Lasse Gerrits[h], Jan Joziasse[i], Vera Maaß [j], Amy Oen[k], Adrian Slob[l], Sue White[m].

[a] Technical University Hamburg Harburg, D-21073 Hamburg, Germany
[b] SEA Environmental Decisions, The Ford, Little Hadham, Hertfordshire SG11 2AT, United Kingdom
[c] Institute of Water and Environment, Cranfield University, Silsoe, Bedfordshire, MK45 4DT, United Kingdom
[d] Cemagref, 69336 LYON CEDEX 9 France
[e] Federal Institute of Hydrology, D-56068 Koblenz, Germany
[f] Institute for Inland Water Management and Waste Water Treatment (RIZA), Ministry of Transport, Public Works and Water Management (RIZA), 8200 AA Lelystad, The Netherlands
[g,l] TNO Environment and Geosciences, 2600 AA Delft, The Netherlands
[h] Erasmus University Rotterdam, Department of Public Administration, 3000 DR Rotterdam, The Netherlands
[i] TNO Environment and Geosciences, 7300 AH Apeldoorn, The Netherlands
[j] Hamburg Port Authoriy,D- 20457 Hamburg, Germany
[k] Department of Environmental Engineering, Norwegian Geotechnical Institute, N-0806 Oslo, Norway.
[m] IWE, Cranfield University, Silsoe, Bedfordshire, MK45 4DT, United Kingdom

1. Introduction

The objective for sediment risk management, as defined by SedNet Working Group on "Risk Management and Communication" , is "to reduce risk posed by contaminated sediments to humans and ecological receptors to a level deemed tolerable by society, and to control and monitor sediment quality and ensure public communication with the final aim of complying with the EU WFD and Habitats Directive". As the term "society" comprises scientists as well as lay people in the field of risk assessment, the level, that is "deemed tolerable", can only be the outcome of a balance of different risk perceptions and priorities. This comprehensive objective has a number of implications:

- Risk assessment is an integral part of, and a tool for, risk management.
- The aim of sediment management is risk reduction; this includes socio-economic, as well as environmental, risks
- While scientific risk assessment can determine the probability of risk and the ecological or human health effects involved, acceptable levels of risk are greatly influenced by society, as well as science.
- To balance the scientific and societal aspects of the decision-making process, meet legal requirements, and obtain stakeholders' agreement, public communication and involvement is an essential part of risk management strategies.
- Any management decision needs to take the time aspects of risk into account, which may last much longer than the actual management process.

Risk reduction at the river basin scale requires a conceptual approach that can be communicated and agreed upon by all stakeholders. The SedNet working group on "Risk Management and Communication" sought to develop such an approach. The steps within this basin scale/site-specific approach include:

- Identification of management objectives
- Determination of appropriate risk indicators
- The use of risk indicators to prioritise sites on a river basin scale and to assess risks on a site-specific scale
- A decision-making process in which potential effects on the river basin and on the site-specific scale are weighed against each other, taking into account the economic, societal and environmental risk, and finally
- Selection, implementation and monitoring of the final management option.

These steps that lead to a final management decision need to be strongly accompanied and guided by communication between all stakeholders.

Sediment management objectives and risk indicators

Identification and understanding of the different management objectives that drive the potential decision making is crucial for opening up communication processes, facilitating the selection of measures or management options, as well as helping to avoid alienation of specific stakeholder groups. The different management objectives on both basin and site-specific scale are:

- meeting regulatory criteria,
- maintaining economic viability,
- ensuring environmental quality and nature development, and
- securing quality of human life.

A comprehensive description and knowledge of the social or societal forces that lead to different management objectives, of the risks that these objectives aim to avoid or reduce, and of indicators that allow monitoring and evaluating these risks, is the prerequisite for any decision on a management options. A possibility to describe and structure the relationship among these different objectives and parameters of risk management is shown in Figure 1.

Risk indicators are measurable attributes that quantify the risk assigned to a situation or an action. They can serve several functions: They can help simplify complex information, can be used for site prioritisation or site-specific risk ranking and can thus trigger management actions. Social and societal forces include <u>human values</u> that have been documented in form of directives and legislation, they comprise <u>human needs</u> that are of concern for society, such as fishing, recreation and the necessity to navigate on waterways for trade purposes, and they represent <u>societal expectations and perceptions</u>, like the perception of risk and the willingness to keep or improve life circumstances.
The **management objectives** aim to minimise risks and negative impacts to the environment; this is done by implementing management options if there is an indication that risks are present (retrospective) or can be expected (proactive). The process of analysing indicators of risk thus links objectives to options. Therefore, selection and evaluation of management options are based upon societal forces, management objectives and risks.
In the conceptual framework of Figure 2, societal forces and management objectives are reflected in the Basin Objectives (BOs).

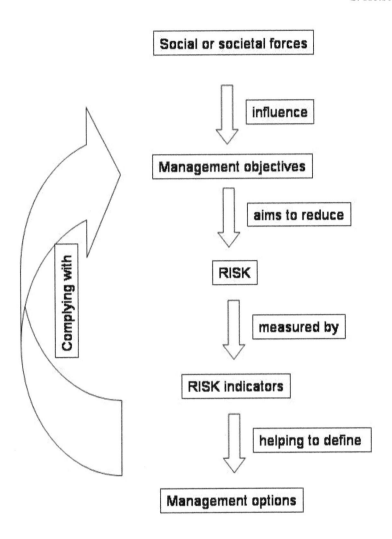

Figure 1: Structural approach of risk management parameters leading to management decisions

Management options can be applied at different scales: risks can be managed on a site-specific, river basin and/or catchment scale. The choice of management options and where to carry them out should be made after consideration of both the site-specific situation and the effect of both current conditions and proposed actions on adjacent sites and the river basin. The evaluation of site-specific and river basin-wide risks is an interactive (and iterative) process, leading to site prioritisation and allowing for allocation of funds to those sites with the highest expected effect on risk reduction in the river basin.

Different indicators of risks can be selected on a site-specific as well as on a river basin scale, addressing water and sediment quality and quantity concerns. Selected indicators can help prioritise sites for further assessment and management, can be used for a site-specific risk ranking of sediments, and, at times, may aid in the selection of management options.

Table 1: Properties of site-specific and river basin risk indicators (RI)

Properties	Site-specific RI	River basin-wide RI
Responsiveness to events	High	Low
Implications for river basin risk	Moderate	Very high
Risk perception of individuals	High	Low
Rate of data gathering	Fast and easy	Complex

Although site-specific and basin-scale risk indicators may be the same (e.g. increase of flood events, decrease of fish yield), they have different properties and may lead to different outcomes (e.g. a local reduction in fish populations may or may not affect the whole river system, depending upon the ecological role of the site).

Strategic Frameworks for Managing Sediment Risk at the River Basin and Site-Specific Scale

Long-term risk reduction in a river basin needs a basin-scale decision framework. In dynamic river basin systems, contaminants and particles can partition, transfer and move through the ecosystem via various media, including air, sediments, soil, water and biota. Thus the management of risk (whether to the environment or to our socio-economic goals) in a river basin, demands that sediment risk management should be closely linked with the assessment and management of these other media. Soil and sediment are present in various depositional areas, and, under the right hydrological conditions, they are amenable to mixing, erosion and further transport downstream. Furthermore, continuing agricultural and industrial practices, as well as catastrophic spills, accidents, and changes in erosional and depositional patterns due to climate change and anthropogenic activities continue to provide both point and diffuse sources of sediment, both contaminated and uncontaminated, into many river basins.

Because of the dynamic, complex and interconnected nature of sediments, from land and rivers to estuaries and the sea, effective and sustainable management strategies must focus on the entire sediment cycle, rather than on one unit of sediment at a time. Such an approach will help focus limited resources to maximise the achievement of management objectives, including basin-scale risk reduction. To achieve this, the relationships between hydrodynamically connected sites, in terms of quality, quantity and energy, must be used to help define their relative risk, their (risk-based) priority in a risk management strategy, how current or proposed management activities may affect one another, and how to best control point and diffuse sources. An integrated, multi-media management framework should be comprised of two principal levels of decision making; the first being a basin-scale evaluation (prioritisation of sites for further evaluation and/or management), the second being an evaluation of specific sites, which includes site-specific risk assessment and, if required, an evaluation of potential management options (comparative risk assessment).

Conceptual Basin Models

The dynamic interactions between sites and matrices (sediment, water, soil, groundwater, air) are complex, and tools are required to better understand and communicate these interactions. Just as a Conceptual Site Model allows risk assessors to consider the flow of contaminants to target organisms in support of site-specific risk assessment, an understanding of the particle and contaminant flows and interactions within a river basin in support of basin-wide management and prioritisation, should inform basin-scale evaluation. This description and inventory, whether conceptual or quantitative, of the mass flow of contaminants and particles within a river basin can be termed a Conceptual Basin Model . It is a critical part of effective risk assessment of a particular site, and of river basin management as a whole. The complexity of a Conceptual Basin Model will differ from river basin to river basin, depending on the information available. It can be quite conceptual, or it can involve detailed chemical, sedimentological, hydrodynamic and modelling studies. Inasmuch as it describes how materials move and interact between sites, it leads to increased knowledge about the river basin system and also serves as an important communication tool about Cause (Drivers-Pressures-State) and Effect (Impact) relationships in a river basin. The more such relationships are understood and communicated, the easier it will be able to manage the system (Response), with full participation of all stakeholders.

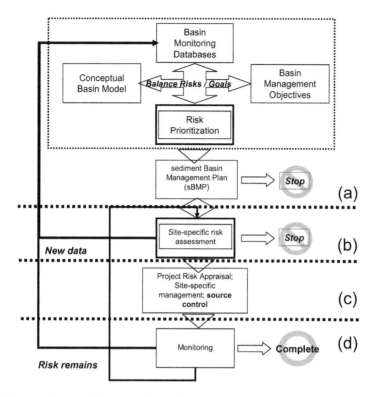

Figure 2: Process diagram for basin-scale (level a) and site-specific risk management (level b). Note: whilst the focus of this figure is on sediment risk management, the same concepts could be applied to other media.

2. From Site Prioritisation (Basin Scale) to Risk Assessment (Local Scale):

If by any process within the river basin sediment at a site poses a risk to one or more of the basin objectives, reduction of that risk may be deemed necessary for effective risk management. In most cases risk reduction is effectively achieved if the whole river basin is addressed because of the dynamic, complex and interconnected nature of sediments. It is important, however, to note that risk assessment is only one aspect that informs the decision process, and the ultimate decision process requires a balance of results from socio-economic, technical, risk and regulatory appraisals. Various aspects of the Conceptual Basin Model should provide screening-level information about how sediment and contaminant distribution and dynamics might possibly endanger the agreed-upon river basin management objectives, as well as how these actions might be addressed. Data from the Conceptual Basin Model, as well as a site risk prioritisation and a consideration of the basin management objectives, will ultimately lead to the inclusion of sediment aspects in a River Basin Management Plan.

Comparing and consequently ranking of sites at basin scale involves setting priorities that account for several management objectives (termed "site prioritisation" in this document), while ranking at site-specific scale is needed to characterize risk for any given management option (termed "risk assessment" in this document). Fig. 3 illustrates the idealised interaction of management processes at basin and local scales: the prioritisation of sites should be done with regard to the whole river basin, and thus must be in the responsibility of the higher (catchment) authority. Once sites have been prioritised, local managers must select final management options on the basis of a site-specific risk ranking, societal values and management objectives (Fig. 3). Any management decision usually requires consideration of more than one management objective.

For basin-scale site prioritisation (i.e. an initial analysis that will determine priority for further study and management) to be affordable, sites should be evaluated at a screening level, with reasonably cost-effective measures that will be consistent with and applicable to further studies. Criteria that should be considered to prioritise sites, and potential measures for their assessment are:

- the location along the up- and downstream gradient
- the potential energy, which could be represented by the slope
- the quantity encompasses two aspects: sediment budget (is there an excess or a deficit of sediment) and mobility (what is the erosion potential and how does it compare with expected and measured shear stresses).
- the quality of the sediment
- the expected benefit if the sediment was to be managed: decrease of related risks, either in place or downstream.

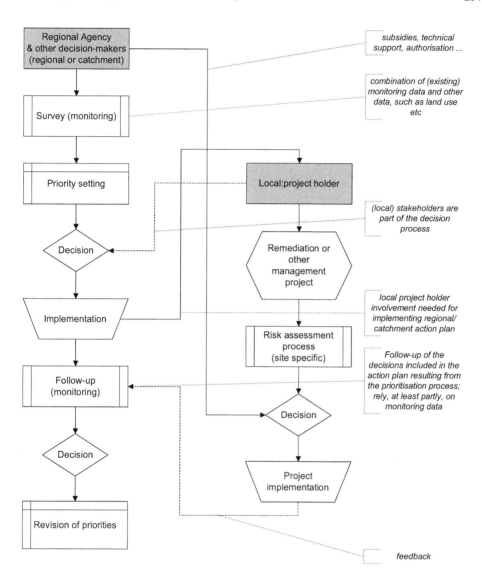

Figure 3. Idealised interaction of management processes at catchment and local scales

Different approaches are possible to classify sites on the basis of these criteria. Two such approaches are score ordination and fuzzy logic. In order to determine the status of each site in the river system for each criterion above, criterion classes must be defined. A simple system would rely upon three classes for each criterion (e.g. high moderate, low). In such a case, two boundary conditions are needed for each criterion, and each criterion must be defined in measurable terms. A higher number of classes can be used, but whether this would improve predictive capability is subject to study. Ideally, these criterion classes should be defined by consensus among experts, including the stakeholders along the river basin, and under consideration of the specific properties of the catchment area.

3. European approaches and regulations

Risk assessment approaches in Europe

If a site is identified as high risk during site prioritisation and development of an sBMP, then it will be subject to a site-specific risk assessment. A site-specific risk assessment is needed in order to determine, in greater detail, what the risks at a given site are. If risks are determined to be unacceptable, or if sites are to be managed to meet socio-economic objectives, the selection of management or disposal options requires a comparative risk assessment that identifies (and possibly compares) the risks to the environment due to those management options. Management options are also driven by site-specific impact on basin management objectives, site-specific risk, technical feasibility, cost/benefit and regulations. Post-remedial monitoring, to confirm risk reduction, flag continuing problems, and to update and refine Conceptual Basin Models is recommended as well.

A proposed conceptual approach to basin-scale sediment management (see Figure 1) provides a framework for addressing the complexities inherent in managing sediments at both a basin-wide and site-specific scale. Acceptance and implementation of a basin-scale approach will require significant work, both technical and political. Successful development of a basin-scale decision framework should provide a basis for parties with very different goals for sediment to come together in support of sustainable sediment management.

In Europe, two major categories of sediment risk assessment have evolved:
- Assessment of the *in situ* risks caused by contaminated sediments in place
- Assessment of risks of contaminants in sediments during and after dredging (*ex situ*)

The first deals with a (site-specific) evaluation of historically contaminated sediments, the second with choosing between management options for the relocation, disposal or treatment of dredged material.

Environmental risk assessment has received much attention in recent decades. As evidence that chemicals released into the environment pose risks to human health and ecosystems grew, many countries have made efforts to control these releases. Most so-called ecological risk assessment frameworks are actually carried out via biological effects-based sediment quality assessments. Most environmental risk assessment frameworks rely on use of multiple lines of evidence (LoE) to infer risk. Important LoEs are: 1) Assessment of predicted or measured sediment toxicity by using chemical measures and bioassays; 2) Assessment of health of the benthic macroinvertebrate community; and 3) Assessment of potential effects of food chain poisoning (evaluation of bioaccumulation and biomagnification). An overview on ecological risk assessment frameworks used in Europe suggests that the term 'decision' means different things in different countries. The information that is collected or available differs, as do the 'depth' of environmental policies and the management objectives of local or national authorities. Hence, a comparison or synthesis of European approaches requires specific attention to the purpose that environmental risk assessment fulfils in each country.

In situ sediment risk assessment
The biological effects-based assessment of the *in situ* risks in sediment focuses on site-specific conditions with respect to the bioavailability of contaminants and the assessment of the damage to the ecosystem. The assessment of damage to the ecosystem can be either predictive or retrospective. If relevant, ecological risk assessment should be combined with studies focusing on human health risks and with studies on the risks related to the transport of contaminants to the surface water or to the groundwater.

With the growing concern over the potential problems caused by sediment contamination, ecological risk assessment approaches have been proposed as decision support tools and as instruments for prioritisation. These approaches generally rely upon a tiered process, to allocate limited technical and financial resources. SedNet suggests an integrated tiered risk assessment approach for deriving decisions of high certainty in a cost-effective manner (e.g. "easy to use" chemical analysis and bio-tests). Assessments used in the first tier should be harmonised along a river basin for comparability of data up- and down-stream and should be applicable as monitoring tools (see discussion below). Results of this first tier should be over-protective because lack of evidence of effects on this first level will remove sediments from further management or analysis, while evidence of effects will trigger a second tier with more sophisticated tests to refine the assumptions and reduce uncertainty. The details of second tier analyses should be tailored based upon site-specific conceptual site models, but should be within a harmonised framework. Parallel to the tiered ecological risk assessment scheme, tiered approaches for human health risk assessment and for risks of transport of contaminants to the surface water or groundwater should be carried out where necessary. In case of human health risks this may involve studies on biomagnification, pathogens and on exposure to contaminats released from the sediment in bathing areas. For a complete assessments of the environmental risks of contaminated sediments, the results of all relevant risk pathways should be assessed in an integrated way at the different levels.

Risk assessment for Dredged Material management (*Ex situ* sediment risk assessment)

Risk assessment for Dredged Material is in fact a *hazard assessment*, in which biological/toxicological endpoints are used as predictors of effects that may occur when the sediment is disposed of in the environment (aquatic or on land); it may also be a risk assessment, provided exposure of the receiving media and ecosystem is accounted for. In risk assessment for dredged material, ideally bioassays are included in the first tier, because they 'widen the window on the chemicals present in the sediment. The approach is more prognostic, i.e. based on the outcome of the assessment, predictions are made of the consequences of disposal of dredged sediments in the environment.

A comparison of *in situ* and *ex situ* risk assessment shows that the assessments lie at very different levels in a decision making process. *In situ* risk assessment is usually carried out as a front-end investigation necessary to evaluate whether sediments pose a risk, before any decision about management actions. *Ex situ* risk assessment is carried out after it has already been proposed to dredge (e.g. dredging for nautical reasons), but when disposal options must be evaluated. It should be pointed out that in those cases where *in situ* risk assessment indicates the need for a management action, the selection of remedial options will require a comparative risk analysis of various management options. However, since the range of such options (including countless variations and combinations of source control and natural recovery; in-place containment or treatment; removal and treatment) is much broader than options for dredged material disposal, the approach to these comparative risk analyses is by necessity highly case-specific. There is a strong need to come up with harmonized assessment methods along river basins where nowadays classifications follow different goals and even the selection of compounds for chemical analysis is not comparable among countries. Because the EU WFD focuses primarily on water quality, it can be expected that *in situ* risk assessment approaches will be used mostly as a diagnostic tool, i.e. to determine whether a poor ecological status of waters is caused by sediment contamination. These approaches may already be part of regulations on a national or regional level, or will be. Harmonisation of these approaches will be difficult, and not desirable because they support location-specific decision making. When the relocation of dredged material in surface water is concerned, there will be more need to harmonise the sediment quality assessment in river basins. The *ex situ* risk assessment can help to more effectively prioritise dredged sediment with high ecological risks, that should be transported to confined disposal sites. Such prioritisation can better be made based on effect observations than on chemical measurements, because biological responses integrate the effects of all biologically available contaminants.

It is assumed, that the Water Framework Directive will have an influence on sediment management practices where the risk for contamination of drinking water and of surface waters by (naturally) suspended contaminated material or relocation of dredged material is concerned. This development should be supported and pushed.

The regulatory situation in Europe with regard to sediments

In the framework of EU legislation there are a few directives directly dealing with sediment or soil, however, none of them concerning directly or exclusively dredged material. Beyond these directives there are a number of Council Directives not specifically addressing sediment or dredged material but which may still concern sediment management.

Among several European countries successful co-operation is practiced in the environmental management of rivers basins. Some of them, e.g. for the rivers Rhine and Elbe, have developed guidance with respect to sediment management, in particular concerning the management of dredged material.

International definitions do not see dredged material a priori as waste. However, the EU Framework Directive on Waste (75/442/EEC, table 3) categorizes sediment dredged from waters as "waste". This means that simply the intention or the requirement to discard sediment after it is dredged turns this material into waste. This view is far away from any hydrological and ecological considerations. It disregards the origin and the role of sediments in the aquatic environment, from water bed stability to the substrate for biota. In addition, this definition, if applied unwisely, would prevent large volumes of clean sediments from being relocated in watercourses with a minimum of ecological impacts and costs. The fact is that dredged material may be, and in reality often is, mainly clean sediment, having no unacceptable ecotoxicological characteristics. This material is still designated and handled as "waste", at unnecessary expense of time and money.

On the other hand, there is no doubt that part of the dredged material is contaminated ("environmentally unacceptable for unrestricted use"). In that case this material must be handled as waste, the potential impacts of relocation and disposal must be assessed, and management options for disposal or treatment have to be developed.

4. Assessment of sediment quality and risks

Chemical and ecotoxicological criteria for sediment quality are an essential component in sediment quality assessment and management. However, at present they exist mainly on national levels, their derivation and application not being harmonized internationally. The exemption is a list of chemical criteria for freshwater sediments developed in the International Commission for the Protection of the Rhine. Chemical quality criteria for coastal sediments proposed by OSPAR are still provisional.

From the regulatory side, use of ecotoxicological data requires elaboration of performance and/or interpretation in order to be able to compare results gained in different laboratories. Interpretation of results for some tests is scientifically complex. Further challenges are the development of a "quality scale" (sediment quality criteria) for ecotoxicological tests which is necessary for the sediment manager to decide how to handle the material, and schemes to deduce decisions from the complexity and often on first sight contradictory results of chemical and ecotoxicological test systems.

5. Stakeholder involvement, risk communication and risk perception

Sediment management will have impact on specific sites, regardless the scale of the problem and the chosen management option. This means that people will be affected by the measures taken and that turns them into stakeholders. Stakeholders are defined as people or organised groups of people who have an effect on or are affected by sediment management. Stakeholder involvement is increasingly becoming part of legislation concerning water bodies in Europe. The communication of the nature and magnitude of the risks of these impacts is one part of stakeholder involvement and a challenge that must be faced if the public is to be involved in the decision making process. Despite a variety of regulations on the need to involve the stakeholders in environmental decision making there is often a low willingness to involve them. Experts and politicians fear that management decisions could be openly rejected, retarding the process and increasing costs. The assumption by these groups is that there is no connection between scientific risk assessment and the public perception of risk. Involvement of stakeholders, therefore, is often lacking.

Understanding how risks are perceived and what governs the public assessment is a first step towards stakeholder involvement and is necessary for decision makers to communicate effectively with the public about sediment. Different stakeholders have different perspectives. This means that they also have different views on risks. Because of these different perceptions of risks, people or organisations (e.g. NGOs) can make different decisions when dealing with the 'same' situation that involves a certain risk. Three different perspectives can be distinguished, based on literature and empirical research. From these three perspectives follow stances towards risks, and possible ways to deal with them. They are summarised in the following table.

Table 2: Language and blind spots for the different perspectives of stakeholders

Perspective	Language	Blind spots
Controller	Danger/Safety (of sediments) Flooding (caused by sediments) People/Society (sediments influencing the quality of life, smell/dirt). Regulation (concerning acceptable waste levels in sediments) Research (on every aspect of sediment) Government (implementation and control of sediment regulation) Control (government has control over sediments and can manage these using regulation	Long term impact (the focus of this perspective is on the medium short term) Alternative solutions (because of the importance of control, some more experimental or riskier solutions will be overlooked)
Guardian	Nature/Ecosystem (sediment management harming flora and fauna) Waste (sediments are contaminated and are damaging the ecosystem) Risk (there is a possible risk that sediments could harm the ecosystem) Damage to the ecosystem by putting polluted dredged material on land.	Economically viable (short term issues are not important, next generations have to be taken into account. Economic viability is therefore not an issue) Efficient/Effectively (nature has to be protected, whether this is costly or not) Short term impact (short term issues are not important, next generations have to be taken into account. Little attention for short term impact of actions)
User	Profit (shipping, sediments are blocking waterways and could endanger profit, sediments should be removed) Resource (sediment as building material or fertilizer) Technology (sediment problems can be solved using technology, treatment) Pragmatic (if sediments cause problems they have to be managed) Costs (the management of sediments can be costly)	Long term impact (short term profits are important) Ecosystem (ecosystem is not an issue because technology can solve all problems, and the ecosystem will restore itself)

It has been observed that the risk tolerance of the stakeholders decreases when access to information is limited and when the public feels powerless and controlled by external forces. For legal, logistical and ethical reasons, complications can only be prevented if the stakeholders are involved in the decision processes from the beginning, and thus the development of risk communication methods is essential.

The different scales of sediment management have consequences for the selection of stakeholders. Site-specific risk communication has to focus on stakeholders on a local level, for example, farmers, companies and citizens that are directly affected by the sediment management on the specific site. On a river basin scale, stakeholders can be on a regional and national level such as representatives of national farmer organisations, branch organisations and environmental organisations. Regardless of the stakeholder group, the perceived risk will be stronger than the scientifically estimated risk.

Communicating with stakeholders is not only about telling the message of the experts and the decision-makers. It is also about involving the stakeholders in the process and about being clear about their role. Stakeholders can use their obstructive power in order to slow down the management process. Inclusion of stakeholders can also mean enrichment of the process.

Potential pitfalls to successful stakeholder involvement include unequal access to information, power or participant number between groups; conflicting expectations; or difficulties with inter-cultural communication. As most people are unaware of any sediment-related issues, experts on sediments should not rush and should take time to explain the issues. Even more, the stakeholders' concerns should be put central. Thus, proper risk communication begins with understanding and respect for people's priorities and a willingness to put issues in those terms. This is probably a valuable lesson for many scientists, who come to meetings with piles of data but little perception of their audience.

The fact that stakeholders can perceive risks in a different way and that they think and act accordingly can make communication and policy-making difficult. It may even be necessary to modify the management options according to the risk perception of some stakeholders. However, insight into risk perception shows that there are ways to cope with non-scientific perspectives. Decision makers must accept, that cooperation is required not only with environmental scientists and engineers but also with sociologists, if they want *"to reduce risk posed by contaminated sediments to humans and ecological receptors to a level, deemed tolerable by society and to control and monitor sediment quality and ensure public communication with the final aim of complying with the EU WFD and habitats directive."*

6. Recommendations

<u>Management objectives and risk indicators</u>

- For the respective river basins a structural approach should be developed in which the social and societal forces, the objectives of risk management and the potential management options are described in their interaction and the prevailing interests are identified.

- By understanding and accepting management objectives of stakeholders, on both the basin and site-specific scale, selection of measures or management options could be facilitated, alienation of specific stakeholder groups avoided and termination of communication prevented.

<u>Development of a river basin scale framework</u>

- Management of risk in a river basin demands that sediment risk management should be closely linked with the management of soil, water, and industrial and agricultural policy.

- An understanding of the particle and contaminant flows and interactions within a river basin should inform basin-scale evaluation. This can be termed a Conceptual Basin Model. It describes how materials move and interact between sites and media, leads to increased knowledge about the river basin system and serves as an important communication tool between scientists, decision makers and stakeholders. Thus, tools in support of Conceptual Basin Model development and use are required.

- An integrated, multi-media management framework should be comprised of two principal levels of decision making; the first being a basin-scale evaluation (prioritisation of sites for further evaluation and/or management), the second being an evaluation of specific sites, which includes site-specific risk assessment and, if required, an evaluation of potential management options (comparative risk assessment).

- Contamination, sediment, and thus related risks, can result from many types of sources. Unless all these risk sources are managed, risks will continue and spread.

- Whilst there is a need (and a requirement by the Water Framework Directive) to stop separating dredging and cleanup, sediments, soil and water in environmental management, there are still significant barriers to be resolved. Conceptual approaches to basin-scale management are proposed, but joined-up policies, uniform datasets and modelling tools still require substantial development.

Site-prioritisation and risk assessment
A prerequisite for sediment management on river basin scale is the harmonization of site -prioritisation and site-specific assessment schemes. This comprises

a) for prioritisation at basin scale
- Development of (an) appropriate indicator(s) for sediment mobility at catchment scale
- Determination of the sediment dynamics and budget in a river
- Development of databases for testing (and improving) prioritisation methods

b) for risk assessment / risk characterisation at local scale
- development of explicit measures of exposure, related to ecological processes which must be done under consideration of management options (or scenarios)
- collection and gathering of data from (local) risk assessment studies (establishment of data bases)
- Harmonization of risk assessment approaches
- Establishment of obligatory monitoring after sediment management (collate data, use them for "validating" effect or exposure assumptions (class boundaries etc))

Establishment of a tiered approach
A tiered approach for risk assessment is to be established for remediation purposes, in which the first tier comprises "easy to use" bio-tests and chemical analysis for risk assessment. This first tier should be harmonized along a river basin for comparability of data up-stream and down-stream and should be applicable as a monitoring tool. Results of this first tier should be over-protective because effects on this first level will then trigger a second tier with more sophisticated test systems in order to refine the assumptions. Parallel to the tiered ecological risk assessment scheme, a tiered human health risk assessment should be carried out where necessary (biomagnification, pathogens and direct exposure in bathing areas). Responses of both lines of evidence should be assessed in an integrated way at the different levels.

Diversity of sediment regulations and monitoring
Assessment of sediment quality and risks: For a basin-wide quality assessment of sediments and subsequent management decisions, harmonised sediment quality criteria and assessment procedures have to be developed.
Regulatory situation: From the scientific-hydrological view, dredged material is primarily natural sediment (eventually contaminated by external sources) while from the political / legal view, it is per se waste. A new view at the role and new

definitions of sediments and dredged material should be developed in the frame of a basin-wide sediment management.

Public communication and risk perception

Respect the risk perception of the stakeholders, even though this does not 'comply' with the scientifically calculated risk

Different stakeholders have different perspectives, or worldviews, meaning that they also have different views on risks. Using a diversity of communication material that uses different words, images and media that are linked to 'triggers' and 'blind spots' of these different worldviews, will help to reach the different stakeholders.

When involving stakeholders in a decision process it should be done at an early stage.

Glossary

Term	Definition	Source
Assessment endpoint	An assessment endpoint describes the effects that drive decision making, such as reduction of key populations or disruption of community structure. An assessment endpoint often has more than one measurement endpoint associated with it,	(1)
Basin scale	see River Basin Scale	
Basin Use Plan	See River Basin Management Plan	
Bioaccumulation	Progressive increase in the amount of a substance in an organism or part of an organism which occurs because the rate of intake exceeds the organism's ability to remove the substance from the body.	(2)
Bioavailability	The capacity of a chemical constituent to be taken up by living organisms either through physical contact or by ingestion.	(3)
Biomagnification	Sequence of processes in an ecosystem by which higher concentrations are attained in organisms at higher trophic levels (at higher levels in the food web); at its simplest, a process leading to a higher concentration of a substance in an organism than in its food.	(2)
Catchment	see River Basin	
Comparative risk assessment	Process that generally uses the judgement of experts to predict effects and set priorities among a wide range of environmental problems.	(4)
Conceptual Basin Model	An understanding of the particle and contaminant mass flows within a river basin in support of basin-wide management and prioritisation. It is the relationship between hydro-dynamically connected sediments, in terms of quality, quantity and energy, that defines their relative risk, and their priority in a risk management strategy	(5)

Conceptual Site Model	The CSM is a three-dimensional description of a site and its environment that represents what is known (or suspected) about the contaminant source area(s), as well as, the physical, chemical, and biological processes that affect contaminant transport from the source(s) through site environmental media to potential environmental receptors. The CSM identifies assumptions used in site characterization, documents the relevant exposure pathways at the site, provides a template to conduct the exposure pathway evaluation and identifies relevant receptors and endpoints for evaluation.	(6)
Contaminated sediment	Sediments that have accumulated hazardous (intrinsic physical / chemical activity) substances (as a result of anthropogenic activities).	(7)
Contamination	Introduction into water, air, and soil of microorganisms, chemicals, toxic substances, wastes, or wastewater in a concentration that makes the medium unfit for its next intended use.	(4)
Disposal	broad term describing all placement of dredged material in the terrestrial or aquatic environment	(8)
Dredged Material	sediments or rocks with associated water, organic matter etc. removed from areas that are normally or regularly covered by water, using dredging or other excavation equipment	(9)
Drivers (of environmental management)	Drivers are the social or societal forces that motivate or otherwise drive sediment management. These include human values that have been documented in form of directives and legislation, they comprise human needs that are of concern for the society, such as fishing, recreation and the ability to navigate on waterways for trade purposes, and they represent societal expectations and perceptions, like the perception of risk and the willingness to keep or improve life circumstances. As in practice, these attributes are represented by the various stakeholders.	Chapter 2, this book
Ecological Risk Assessment	The application of a formal framework, analytical process, or model to estimate the effects of human actions(s) on a natural resource and to interpret the significance of those effects in light of the uncertainties identified in each component of the assessment process. Such analysis includes initial hazard identification, exposure and dose-response assessments, and risk characterization.	(4)

Environmental / Ecological Risk	The potential for adverse effects on living organisms associated with pollution of the environment by effluents, emissions, wastes, or accidental chemical releases; energy use; or the depletion of natural resources.	(4)
Exposure	The amount of radiation or pollutant present in a given environment that represents a potential health threat to living organisms.	(4)
Exposure assessment	Identifying the pathways by which toxicants may reach individuals, estimating how much of a chemical an individual is likely to be exposed to, and estimating the number likely to be exposed.	(4)
Hazard	the inherent toxicity of a compound. Hazard identification of a given substances is an informed judgment based on verifiable toxicity data from animal models or human studies.	(4)
Human Health risk	The likelihood that a given exposure or series of exposures may have damaged or will damage the health of individuals.	(4)
Management objective	Management objectives represent the overall aims that direct management options and are governed by drivers. Objectives in sediment management are the need to meet regulatory criteria, to maintain economic viability, to ensure environmental quality and nature development, and secure quality of human life.	Chapter 2, this book
Measurement endpoint	Measurement endpoints approximate, represent, or lead to the assessment endpoint, using field or laboratorymethods.	(1)
Monitoring	Periodic or continuous surveillance or testing to determine the level of compliance with statutory requirements and/or pollutant levels in various media or in humans, plants, and animals.	(4)
Relocation	Relocation of dredged material is a coherent physical and technical process of dredging, transporting and aquatic disposing of dredged material in a nearby place within the same waterbody.	(8)
Risk	A measure of the probability that damage to life, health, property, and/or the environment will occur as a result of a given hazard.	(4)
Risk analysis	risk analysis can be described as a scientific approach towards risk, used for public policy making on technological, environmental and health issues.	(10)

risk assessment	The procedure in which the risks posed by inherent hazards involved in processes or situations are estimated either quantitatively or qualitatively.	(11)
Risk characterisation	Integration of evidence, reasoning, and conclusions collected in hazard identification, dose-response assessment, and exposure assessment and the estimation of the probability, including attendant uncertainties, of occurrence of an adverse effect if an agent is administered, taken, or absorbed by a particular organism or population. It is the last step of risk assessment.	(2)
Risk communication	Interactive exchange of information about risks between stakeholders.	(2)
Risk indicator	risk indicators in the scope of this book are measurable parameters that trigger for choosing management options on site-specific basis, considering the effects on and the scope of the river basin.	Chapter 2, this book
Risk management	Process of evaluating alternative regulatory and non-regulatory responses to risk and selecting among them. The selection process necessarily requires the consideration of legal, economic and social factors.	(12)
Risk perception	The perception of a natural person or stakeholder of expected losses (of lives, persons injured, property damaged and economic activity disrupted) due to a particular hazard for a given area and reference period.	(12)
Risk ranking	Evaluating individual sediment parcels to determine and rank their risk relative to benchmarks, site- or basinspecific criteria.	(5)
River Basin	River basin. means the area of land from which all surface run-off flows through a sequence of streams, rivers and, possibly, lakes into the sea at a single river mouth, estuary or delta.	(13)
River Basin Management Plan	River Basin Mangement Plans, as required by the Water Framework Directive, are considered to be management tools to achieve the environmental objectives of the WFD. They should comprise among other issues a general description of the characteristics of the river basin district, a summary of significant pressures and impact of human activity on the status of surface water and groundwater, a list of the environmental objectives, and a summary of the programme or programmes of measures.	(13)

Sediment	suspended or deposited solids, acting as a main component of a matrix which has been or is susceptible to being transported by water.	(14)
Sediment assessment	the process used to characterize sediment for a given purpose (e.g., evaluations for risks to environmental health, dredged disposal, land farming, habitat construction, etc.).	(15)
Sediment management	the process of making decisions and taking actions on sediments, taking into consideration a wide range of factors.	(15)
Sediment management strategies or options	the range of actions that can be taken once risks have been assessed and risk managers have balanced those risks against various objectives and goals. These range from no action (either because risks do not exist or are not controllable) or institutional controls to more aggressive containments, treatments or removal actions.	(15)
Sediment risk management objective	The sediment risk management objective, as it has been agreed upon during the first SedNet Workshop of Working Group 5, was to reduce risk posed by contaminated sediments to humans and ecological receptors to a level, deemed tolerable by society and to control and monitor sediment quality and ensure public communication with the final aim of complying with the EU WFD and habitats directive.	(16)
Site prioritization	Evaluating parcels of sediment within a region in terms of hydrodynamics, risk, regulatory and socio-economic goals to rank and prioritise sites for management order or focus.	(5)
Stakeholder	People or organised groups of people who have an effect on or are affected by sediment management	(6)
Sub-Basin	This describes the area of land from which all surface run-off flows through a series of streams, rivers and, possibly, lakes to a particular point in a water course (normally a lake or a river confluence).	(13)
Sub-catchment	see Sub-Basin	
Weight of Evidence Approach	The weight-of-evidence approach is the process by which measurement endpoint(s) are related to an assessment endpoint to evaluate if there is a significant risk of harm to the environment. The approach is planned and initiated at the Problem Formulation Stage and results are integrated at the Risk Characterization Stage.	(17)

Sources

1. Glenn W. Suter II, "Ecological Endpoints," Chapter 2 in USEPA, *Ecological Assessment of Hazardous Waste Sites: A Field and LaboratoryReference* (EPA/600/3-89/013).

2. International Union of Pure And Applied Chemistry (IUPAC) (1993): Glossary For Chemists Of Terms Used In Toxicology: Pure and Applied Chemistry, v. 65, no. 9, p. 2003-2122 (on-line version posted by the U.S. National Library of Medicine).

3. USGS National Water-Quality Assessment Glossary. http://water.usgs.gov/nawqa/glos.html

4. US EPA: Terms of Environment. http://www.epa.gov/OCEPAterms/

5. Apitz, S. and S. White (2003). "A conceptual framework for river-basin-scale sediment management." JSS - J Soils & Sediments 3(3): 125-220.

6. Apitz et al. (2002): Critical Issues for Contaminated Sediment Management. MESO-02-TM-01, March 2002. http://meso.spawar.navy.mil/Docs/MESO-02-TM-01.pdf

7. Protocol of the first workshop of SedNet Working Group „Risk Management and Communication"

8. Dutch-German Exchange (DGE) on Dredged Material - Part 1 - Dredged Material and Legislation. April 2003

9. OSPAR Convention for the Protection of the Marine Environment of the North-East Atlantic. Ministerial Meeting of the OSPAR Commission. SINTRA: 22 - 23 JULY 1998: OSPAR Guidelines for the Management of Dredged Material.(Reference Number: 1998-20)

10. Asselt, M. B. A. van (2000), 'Perspectives on certainty and risk. The PRIMA approach to decision support', Dordrecht: Kluwer Academic Publishers

11. EEA. http://service.eea.eu.int/envirowindows/chapter1h.shtml

12. United Nations. Glossary of environment statistics. http://esa.un.org/unsd/envmnt/default.asp

13. Water Framework Directive, 2000/60/EC

14. Salomons, W. and J. Brils, Eds. (2004). Contaminated Sediments in European River Basins. European Sediment Research Network SedNet. EC Contract No. EVKI-CT-2001-20002, Key Action 1.4.1 Abatement of Water Pollution from Contaminated Land, Landfills and Sediments., TNO Den Helder/The Netherlands. 80p.

15. Apitz, S. E. and E. A. Power (2002). "From Risk Assessment to Sediment Management." J Soils and Sediments 2(2): 61-66

16. Heise et al. (2004). "Risk Management of Sediments and Communication - Synthesis of the SedNet working group 5 outcomes." JSS - J Soils & Sediments 4(4): 233-235.

17. Menzie et al. 1996. Special report of the Massachusetts Weight-of-Evidence Workgroup: a weight of evidence approach for evaluating ecological risks. Human Ecol Risk Assess 2: 277-304

Index